ANCIENT
ALIENS
ON
MARS

MIKE BARA

Adventures Unlimited Press

Ancient Aliens on Mars

by Mike Bara

Copyright 2013

ISBN 13: 978-1-935487-89-0

Published by:
Adventures Unlimited Press
One Adventure Place
Kempton, Illinois 60946 USA
auphq@frontiernet.net

www.AdventuresUnlimitedPress.com

ANCIENT ALIENS ON MARS

Adventures Unlimited Press

Acknowledgements

I would like to acknowledge these special souls who helped,
loved and/or supported me during the writing of this book:
Denise Zak, Shana Paredes, Sara Vasquez, Steven Jones, Derrell
Sims, Maureen Elsberry, Simon Rockell, Ben Fox, Sherri Gaston,
my publisher David Hatcher Childress, Alan Pezzuto, my brother
Dave, Bailey & Barkley, Sebastian and my little lights The Lady
Aurora and Miss Fluffy-Muffy.

TABLE OF CONTENTS

Dedication

This book is dedicated to my friend and former co-author,
without whom the many mysteries of Mars would have
remained forever buried in the sands of time.

See Mike Bara at:

MikeBara.Blogspot.com

FOREWORD

Before we move on to Mars, there are a couple of loose ends to clear up from my previous book published by Adventures Unlimited, *Ancient Aliens on the Moon* (AAOTM). Because of time and space constraints, this new information didn't get published at that time but I feel it is important enough to include here.

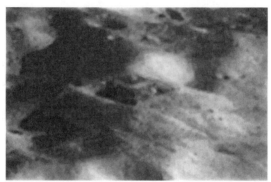

The Daedalus Ziggurat on the far side of the Moon.

Just before the completion of AAOTM, I came across an image of an object which stirred great controversy amongst the debunker elite. But it caused even more angst amongst the "skeptical" small-time bottom feeders. One of these was a man named Stuart Robbins, who I mentioned in the last chapter of AAOTM. After I responded to his absurd accusation that either I or Richard Hoagland had fabricated the Daedalus Ziggurat image, he went out of his way to answer back on his blog.

As usual, he made a lot of assertions, many of which are false, most of which are misleading, and some of which are just plain deceptive. As I skimmed his detailed collection of claims and statements, some of them backed by actual math, I debated not responding at all. But then I noticed at the bottom, where he made it a special point of emphasis to say that he was "Dr. Robbins" not "Mr. Robbins," and I got curious as to who he really was. Because of his association with an obsessive nutcase who has cyber-stalked me and Richard for more than six years, I assumed that he was just another "useful idiot" of the type that NASA depends on to keep the lid on what's really on Mars and the Moon. But then I got curious as to what his doctorate might be in (I was thinking maybe forestry) so I started to look at his personal information. And then I noticed something—his education was funded by NASA.

According to his own website,[1] he got a grant from NASA for his post-doctoral work. He also continues to get funding from NASA[2] for his other research projects, most of which include studying craters. I posted on Twitter about it:

 Mike Bara @mikebara33   8 Aug
Dear Dr. Robbins; I apologize. I thought you were just another dumbass like expat. I had no idea you were actually a paid NASA shill. My bad
Expand

Stuart responded on his blog on August 8th, 2012,[3] admitting that this is accurate—he does indeed get funding from NASA—but downplaying it of course:

> I make a meager living like most scientists and, like most astronomers, a fair amount of my salary does come from NASA-awarded grants, but I literally have less connection with NASA than a custodian who sweeps the floors of JPL.
> – Dr. Stuart Robbins

Yeah, no connection at all, except for the part where they *pay you*… You can no more be a "little bit" on the take from NASA than you can be the proverbial "a little bit pregnant," Stuart.

Now to be clear, I'm not implying that he is taking money directly from NASA ("hush money," he called it) to post attacks against Richard and I on his blog. Although, to use his own phrasing, I "wouldn't put it past him." But being financially dependent on the very institution that Mr. Hoagland and me have challenged and exposed on a regular basis for more than a decade and a half for his rent, food and car payments *by definition* creates a pernicious bias that cannot be overcome. It is an inherent conflict of interest, and it permeates everything he does and writes about us. How can it not?

This changed everything for me. He wasn't attacking me and accusing Hoagland and me of fraud because he was just a psycho, like most of the other critics, he was doing it because he was a *paid shill for NASA*. In fact, it wouldn't surprise me if he was able to respond so quickly and extensively to my posts because he was writing on his personal blog using taxpayer or university funded equipment and internet access while he was supposed to be working. (Note: This suspicion was confirmed when Dr. Robbins posted an update on his personal blog on 8/16/2012 at 3:45 PM in the afternoon,[4] the middle of the work day. It obviously must have taken him at least a couple of hours to write up. I'm wondering which government funded project you charged these hours of work on your personal blog to, Stuart?)

As my dad used to say, "It's good work if you can get it."

At any rate, now that I knew he was *being paid by NASA* to attack me and Mr. Hoagland, I calmed down a bit. After all, like James Oberg and "Dr. Phil" Plait before him, NASA has constantly trotted out one paid shill after another to distort our claims, spread disinformation about us and generally charge us with one nefarious deed or another. It's old hat, and the fact that they are on the NASA payroll completely discredits the shopworn "I'm just an independent skeptic defending the people against pseudoscience" line.

Dr. Stuart Robbins

That of course doesn't stop Sheldon from using it.

Sorry, I mean Stuart.

The fact is, no one who is taking money from NASA, and therefore financially dependent *on* NASA, has any kind of credibility as a "skeptic." A true skeptic is someone who reserves judgment and questions established orthodoxies, paradigms and dogmas. I for instance, was initially skeptical of the Daedalus Ziggurat image, but inclined to lean toward its authenticity because of a variety of reasons I've already stated and will cover in this Foreword. I am also skeptical of NASA's honesty and the integrity of the data they present, due to years of catching them fabricating data and painting over things they don't want the public to see on images from all over the solar system. I also am fully convinced that the official NASA version of Apollo photo AS11-38-5564 has been deliberately altered by NASA to obscure not only the Ziggurat, but a lot of other artifacts all over that image (see AAOTM). I will provide further proof of that later in this Foreword. But Stuart, like Oberg, Plait, Sagan and a whole gaggle of others before him, is not a "skeptic." He is a professional, paid debunker. He is not interested in the truth, and he will never admit to anything that would cast NASA in a bad light. If he did, it would be career suicide.

So let's keep the issue of his credibility as an independent voice out of this. He's on the take, plain and simple. He can no more do a fair and independent analysis of this or any other claim made by me or Mr. Hoagland than Dr. Sheldon Cooper can sit on a spot on the couch other than "his" spot. It's just not in his DNA.

Or his wallet.

So now the point was, why respond at all? I was leaning against doing so, and several close friends (including Richard Hoagland) urged me to put my energies elsewhere. But then, the usual village idiots, a small but vocal collection of Bara-haters who attack virtually everything I post, started coming on to my Facebook page and attacking me, my family and friends in the most personal and vicious manner. Many of them were fake Facebook profiles, created just so they could come in and post nasty stuff about me. They also personally attacked my brother and sent sexually harassing messages about me to several of my more comely female Facebook friends. So I got pissed off.

I responded on my own blog[5] on August 23rd, 2012. Stuart then responded in kind by laying out three pillars of evidence that would need to be discredited before he would change his position that either I or Mr. Hoagland had fabricated the Daedalus Ziggurat image.

(Quick note: to save time on the rest of this Foreword, "Stuart (paid shill for NASA)" will now be abbreviated to "Stuart (PS4NASA)."

From Stuart's (PS4NASA) blog post: "As promised, Mike Bara has posted a rebuttal to my analysis of the lunar ziggurat. To recap from earlier, I noted these three points of what Mike must explain before I would revise my conclusion:"

Well first off, I couldn't care less about what Stuart (PS4NASA) wanted me to "explain" before he "revised" his conclusion. I already knew what his conclusion was going to be, regardless of anything I might or might not explain to him. He is a semi-professional debunker, plain and simple. He was only pretending to be open to "revising" his conclusion to maintain a phony air of impartiality. That said, Stuart (PS4NASA) then started his rebuttal with his "big 3" questions. These three questions are the main pillars of his conclusion that the Ziggurat image I posted is a Photoshop fraud made *from* the official NASA image "5564.jpg." As I will show, all three of them are based on bad data, false claims and incorrect reasoning.

They are, in order:

"1. Why there is less noise in the NASA original but more noise in Mike's, and why is there more contrast (more pure black and more saturated highlights) in Mike's? Both of these pretty much always indicate that the one with more noise and more contrast is a later generation ... you can't just Photoshop in more detail like that."

"2. Why other images of the same place taken by several different craft (including non-NASA ones), including images at almost 100x the original resolution of the Apollo photo, don't show the feature."

"3. Why the shadowed parts of his ziggurat are lit up when they're in shadow, on top of a hill, and not facing anything that should reflect light at them?"

11

Pillar #1 – Noise and Contrast

The first of Stuart's (PS4NASA) 3 pillars of support for his claim that the Daedalus Ziggurat is a hoax is the presence of what he claims is "noise" in the image, "as1120pyramid20smallue2.jpg."

To start with, I initially had no idea why there were differences between the so-called "NASA original"—5564.jpg[6]—normally found on the Lunar and Planetary Institute's "Apollo Image Atlas" and "as1120pyramid20smallue2.jpg." Further, as I have repeatedly pointed out, the Ziggurat image "as1120pyramid20smallue2.jpg" is not "mine," although I have several thoughts on its origins which I will cover in this Foreword. I assume that "as1120pyramid20smallue2.jpg" is different from the NASA image because somebody scanned it, enhanced it, and then reduced it for uploading to the web, but I really wasn't sure of this until recently.

Second, let's look more closely into what Stuart (PS4NASA) is implying in his first challenge to the authenticity of the Ziggurat image.

Stuart (PS4NASA) is suggesting here that "as1120pyramid20smallue2.jpg" is actually made *from* the NASA file "5564.jpg." That's what all his talk about it being a "later generation" is all about. For various reasons, I don't agree with that at all. I think "as1120pyramid20smallue2.jpg" is a completely different file, although scanned from a similar analog photographic print or negative.

But even if it *was* made from "5564.jpg," his arguments in support of this are completely specious. There are NOT, as he claims, "more pure black and more saturated highlights" in "as1120pyramid20smallue2.jpg." Quite the opposite, as I proved in the last chapter of *Ancient Aliens on the Moon,* there is far more pure black in the NASA version because of all the Photoshop paintbrushing they did. What this means is that by his own logic, the official NASA image is the "later generation," and therefore more likely to have been altered or outright faked.

As for the statement "…*you can't just Photoshop in more detail like that,"* I'm not really positive, but I think he's saying that because the faked official NASA image (5564.jpg) has "more detail" in it, that is more proof it must be older or "better" than the Ziggurat image because you can't use Photoshop to add in detail. Therefore by his (incorrect) deduction, the Ziggurat image must have been made *from* the NASA image "5564.jpg." I guess he thinks this is important because he's trying to prove that the Ziggurat image was somehow conjured up from the image currently on the NASA website. And of course he previously implied that either Hoagy or I had done this. But his argument really proves no such thing, even if he was right. All it proves is that one image has more "noise" in it, which by itself is proof of nothing.

Further, his argument about detail actually supports the authenticity of "as1120pyramid20smallue2.jpg" over "5564.jpg" when you dig into the, ahem, details...

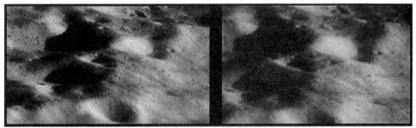

A tale of two Ziggurats: "5564.jpg." (L) and "as1120pyramid20smallue2.jpg." (R)

Looking at the two images side-by-side, it becomes obvious fairly quickly that "as1120pyramid20smallue2.jpg" actually contains considerably *more* detail that "5564.jpg." Unlike the official NASA version, "as1120pyramid20smallue2.jpg" contains subtle greyscale gradations, shadows and details, even in the most darkly shadowed areas of the craters. By contrast, as I showed in chapter 9 of *Ancient Aliens on the Moon*, NASA's "5564.jpg" has no detail whatsoever in its pitch-black craters and sky. All of the pixels are color #1, absolute black, which even people at Adobe (makers of Photoshop) state is a sure sign that the areas have been painted over with some type of paintbrush tool.[7] In other words, these areas show no detail whatsoever, whereas the same areas in "as1120pyramid20smallue2.jpg" *do*. This proves that, using his own standards and contrary to Stuart's (PS4NASA) published assertions, it is the *official NASA version* of the image that has been tampered with, NOT "as1120pyramid20smallue2.jpg."

Which brings us to the questions of "noise," and image enlargement and reduction.

In his post, Stuart (PS4NASA) claims, as others also have, that reducing an image as Stuart (PS4NASA) did when he shrank the Ziggurat image by 15% "reduces noise" and thereby makes the image somehow "better." (I mean after all, if it doesn't make the image "better," why would you do it?)

But once again this is simply wrong, and also counterintuitive. Reducing an image doesn't make it better in any way, shape or form. All it does is reduce the amount of *information* in that digital image.

Here's the original full size image of the Ziggurat that I sent to Richard. Stuart (PS4NASA) took this image and reduced it to 85% of its original size. Now let's take this image of the Ziggurat and reduce it as he did.

I reduced it to 10% to make a point. I'm sure any "normal" person (a word Stuart used when talking about me) would agree that it's worse, not better, than the original. This is because reducing a digital image doesn't just reduce *noise*, it reduces the *signal and information across the board in an image, noise included,* and

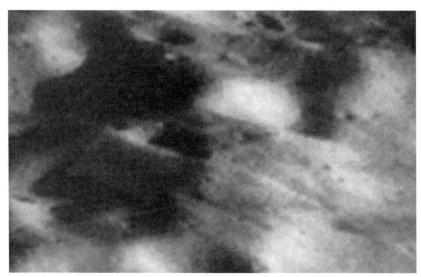

as1120pyramid20smallue2.jpg

makes it more pixelated and less accurate. In short, it reduces the signal AND the noise.

Which is of course, is why he did it in the first place.

In addition, there is also a nifty tool in Photoshop called "Reduce Noise" that does a perfectly excellent job of reducing noise in an image. Here's a version of the Ziggurat that I used the filter on. It's a bit blurry, but you can still see all the major features and it's not reduced to the size of a postage stamp in the process:

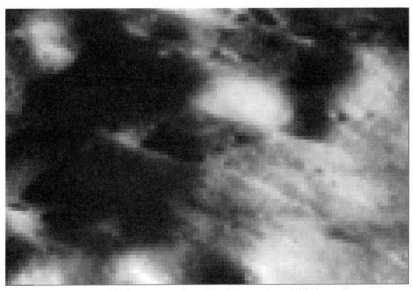

as1120pyramid20smallue2.jpg reduced 90%

Obviously, this version is much better than the Stuart (PS4NASA) reduced version. So any "normal" person *reducing* an image to make it "better" is nonsensical. It can only make it worse, and make fine detail harder to see. Maybe Stuart (PS4NASA) just hasn't noticed that tool in Photoshop in his "20-plus years" of experience working with the program.

Then, Stuart (PS4NASA) tried to justify his actions in a new blog post, and along the way admitted that what I just demonstrated is true:

"Yes, it will reduce some detail. That is true."

15

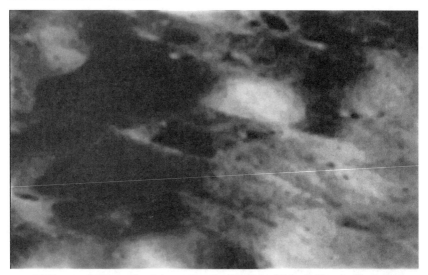

as1120pyramid20smallue2.jpg using the "Reduce Noise" filter in Photoshop

Thanks Stuart (PS4NASA), I knew that already. He then of course goes on to argue that it doesn't really count:

"But at 85.28%, it will not change the detail enough to say "oh look, there's a pyramid there" versus "what happened to the pyramid?!" and it WILL REDUCE the noise by roughly 8ish%."

OK Stuart (PS4NASA), if that's true, then why do it AT ALL? As I just showed you, you can get much better noise reduction with far more precise control by using the "Reduce Noise..." filter in Photoshop. Again, the only reason a "normal" person would reduce an image is to *reduce the amount of detailed information* in the image. Further, his claim that such an action would not reduce the detail enough to make a difference, I guess we just disagree on that. I'm looking at very fine details in the Ziggurat image and I don't want to destroy any of them by reducing it.

Stuart (PS4NASA) then went on to claim that a statement I made in my rebuttal to the effect that interpolating an image will *improve* it is "factually and demonstrably false":

"But I'm sorry, Mike, your statement that upsampling (interpolating) makes an image better is factually and demonstrably false. You cannot get more information than was there originally."

Again, this is a standard debunkers shtick trick. I *never said* that interpolating a digital image can "get more information than was there originally." What I said was: "This upsampling process would have the effect of actually making the NASA image *better*, rather than making the original enhancement *worse*."

Again, this is a standard professional debunkers technique of making me defend/rebut statements I never made. But nonetheless, we'll have a chance to examine this question more closely later in this book, when we get to the Face on Mars. For now, I want to give you just one example proving I'm right and he's wrong.

Here's an image taken from the *Mars Reconnaissance Orbiter's* HiRise camera showing the *Curiosity* rover descending over the Martian sands on its way to a landing in Gale crater. As most "normal" people can see, it is small and rather "contrasty."[8]

Full size image PIA15980-full_full.jpg (NASA)

But a few days later, NASA released a new close-up full size image of the rover and its parachute. Only this one was upsampled (interpolated), contrast enhanced and worked over with specialized filters to remove the noise and sharpen the image.[9]

17

Full size image (NASA)

And a few days later, private enthusiasts did an even better job by interpolating it even more.[10] At this point, I'll leave it to the "normal" people reading this to deduce which image is "better."

So the bottom line is this: Stuart (PS4NASA) didn't reduce the image of the Ziggurat to *improve* it, he reduced it to make it worse, *because that's what data reduction does*. Still, Stuart (PS4NASA) is correct that interpolation doesn't "add" information to an image. But I never said it did. What interpolation does is *enhance* the information that is already there, to make an image *better*. That's why they call it "image enhancement, and not "image degradation."

Before I move on, I want to say one more thing about the "noise" that Stuart (PS4NASA) is so obsessed about. He seems to have seized on this as some sort of proof that the original Ziggurat image has been tampered with. Not only do I dispute this line of reasoning in its entirety—as I have made clear in my blog posts and this Foreword—there is another reason the so-called "noise" doesn't bother me.

I don't think it's noise.

In looking at scans of actual first-generation Apollo photographs, which I have done plenty of, you can see lots of dust and dirt accumulation on the prints, which is especially visible in the dark areas. Now, this may look like noise to a scanner and to Photoshop, but it is really just a normal buildup of residue on the print from 40 years of sitting in a photo album under a plastic sheet.

Photo albums from the 1960s and 1970s had pages with a sticky adhesive on them which was distributed in a very even, spotted pattern, and had thin plastic sheets to protect the prints. But, if a photo were placed under one of these sheets and the opposite page didn't have a photo, then the pattern of sticky spots on the opposite page would eventually leave an imprint the photo, especially if the albums were stacked one on top of the other for oh, say 40 years.

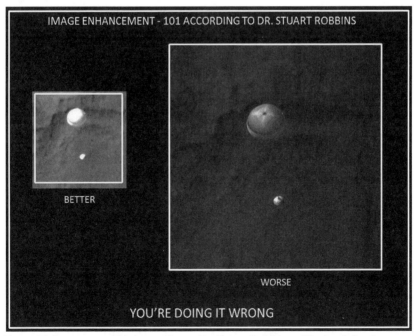

Image Enhancement 101 According to Dr. Stuart Robbins – You're doing it wrong

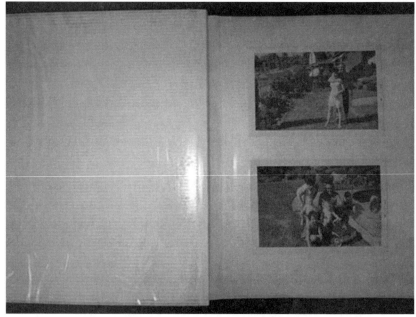

Typical family photo album from the 1960s/70s

Pull any old family photo from a photo album stored this way and you will see this pattern, regardless of how well you've taken care of them. The plastic sheets themselves, pressed tightly against the prints for decades by the weight of the rest of the album and other albums that may be stacked on top of them, will adhere to the print in some places and leave marks on the photo print in the pattern of the little raised sticky spots. If you were to pull this print from the album 30 or 40 years later, scan it at high resolution and increase the gamma, it would show this pattern of marks all over the scan. Such a pattern would have to be removed with a noise reduction filter.

So what I have always suspected is that the so-called "noise" in as1120pyramid20smallue2.jpg isn't noise at all. What I speculate happened is that some NASA veteran had an original or near-first generation print of AS11-38-5564 in his collection, stored it in a photo album under the stairs, and his curious son or nephew came along one day, went through his old photo albums, saw the Ziggurat and said "Holy S***!" This person then scanned and processed the image as best he could and posted it on the web, where it has

Close up of the "sticky dots" glue pattern on a photo album sheet

been making the rounds for a while until I spotted it and gave it to Richard.

Look again at this contrast enhanced version of as1120pyramid20smallue2.jpg and note the sticky spot pattern all across the image. It is aligned with the actual vertical/horizontal of a full print of AS11-48-5564, rather than this rotated close-up. Now go pull an old, pressed down image from a photo album. They'll

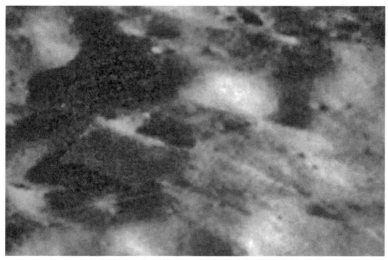
Contrast enhanced version of as1120pyramid20smallue2.jpg showing spotting pattern consistent with long term storage in an old-fashioned photo album.

WHEN IS "NOISE" NOT NOISE?

WHEN IT'S RESIDUE FROM A PHOTO ALBUM SHEET

as1120pyramid20smallue2.jpg rotated so that vertical is "up." Note the pattern exactly matches in alignment and scale with standard photo-album glue spots from the 1960s and 70s.

look exactly the same.

When you rotate the image to match the orientation of the official NASA version, it becomes even more obvious that most of the "noise" Stuart (PS4NASA) claims is on the image is actually photo-album residue. Again, pull some old photos from your family's photo albums and make a comparison. Depending on how long they've been in there, you'll find that they are most likely a perfect match for the residue pattern seen on as1120pyramid20smallue2.jpg. Not only does the pattern match in alignment with the photo album glue residue, it also matches it *in scale*, making it a virtual certainty that is the source of the so-called "noise" on the image.

And again, when you show as1120pyramid20smallue2.jpg and the photo album glue page aligned as they truly would be if the Ziggurat image came from a photographic original stored for several decades as I've described, they match perfectly.

I would place the likelihood of this scenario at about 95%, and this discussion is part of the "due diligence" that Richard

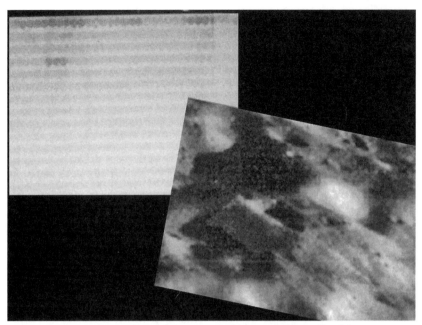

Photo album glue residue on as1120pyramid20smallue2.jpg

Hoagland mentioned on *Coast to Coast AM* the night he revealed the image to the world. Stuart Robbins (PS4NASA) mocked this very "due diligence" on his blog. He decided that the glue pattern was digital noise deliberately induced to cover up tampering with as1120pyramid20smallue2.jpg. As we have proven, it is Stuart (PS4NASA) who failed to do his due diligence, and now he who looks the fool.

What this all means is that Stuart's (PS4NASA) belief that the presence of this glue spot pattern somehow proves that "as1120pyramid20smallue2.jpg" is a later generation image than NASA's "5564.jpg" is false, or at least dubious in the extreme. What then follows is that his further conclusion/declaration that the Ziggurat on "as1120pyramid20smallue2.jpg" must have been "drawn-in" on top of "5564.jpg" is also most likely an incorrect conclusion based on his faulty reasoning.

In fact it is the other way around. The scan of "as1120pyramid20smallue2.jpg" was almost certainly made from an earlier generation NASA print, and "5564.jpg" was made much later – decades later in fact – and the offending Ziggurat was most certainly digitally removed since "as1120pyramid20smallue2.jpg" had already been making the rounds on the web. I would also remind the reader that in Ancient Aliens on the Moon, I already established as a fact that "5564.jpg" had been digitally altered with a paintbrush tool of some kind.

And that pretty much blows point #1 out of the water.

But wait, there's more!

In June of 2013, one of Stuart's sycophants on the web went searching for the origin of as1120pyramid20smallue2.jpg in a vain effort to prove that I had somehow lied about its origins. In the course of doing so, he convinced himself that he had "caught" me in a lie/mistake, and eagerly sent an email to my manager gloating over it. In fact, what he discovered only reinforced my arguments that as1120pyramid20smallue2.jpg came from a scan of an analog print which had been in this person's family for years, and in the process he managed to completely throw Stuart (PS4NASA) under the bus and prove him wrong.

Using the internet archive tool, this troubled individual (he sends me emails on an almost daily basis accusing me of being a "liar" and various other things) went to an archived website of an early anomaly hunter named Terry James, aka "KK Samurai." Now, I knew of "KK Samurai" from the late 90's and frequently enjoyed his finds and articles. When I had first seen the Daedalus Ziggurat while doing research for Ancient Aliens on the Moon, I thought it looked familiar but couldn't place it. It was suggested that the original source might have been "KK Samurai," but I expressed doubts about this because he always watermarked his discoveries with a "KK" symbol in the lower right hand corner, and "as1120pyramid20smallue2.jpg did not have such a watermark. This sycophant (and his sycophants) also mocked Terry James as a "known hoaxer" (which I knew he wasn't), primarily because he was "a Christian." But when they thought they "had" something on me, they withdrew these charges quickly.

What they found were some images on Terry James' archived website that absolutely verified what as1120pyramid20smallue2.jpg showed, and appeared to be made from the same source file. I

Image retrieved from the Internet Archive showing the Daedalus Ziggurat, first posted in 1999-2000. Note "KK" watermark in lower right.

went in and pulled down the images myself to make sure there was no funny business with altering of the images by the sycophant.

In looking over the archived website I saw that in addition to first posting the image, Terry James also had done some colorization work on it. He also posted images showing that author Richard Coombs had made an initial comparison to the Ziggurat at Ur in Iraq, identical to the comparison I made years later in *Ancient Aliens on the Moon* without even knowing about Terry's pages or Richard's analysis.

But the most critical piece of information from the website came from 1999, where Terry James thanked "Frank," for giving him the image. In fact, in the email from the sycophant, he identified the source of the image as a man named Frank Gault:

"The actual source of the image was a scan done by Frank Gault, which is why in the original presentation you see 'Thanks to Frank....' Gault's father was ex-NASA, and gave his son a large collection of Apollo-era 10x8 photo-prints, perhaps similar to Ken Johnston's collection..."

After a quick perusal of the archived KK Samurai website, I was able to confirm most of this information. This sycophant apparently thought he "got" me because of my previously expressed doubts that Terry James was the source of the image (as1120pyramid20smallue2. jpg) I had originally given to Richard Hoagland. In fact, what his digging actually does is to throw Stuart (PS4NASA) Robbins completely under the bus.

As I established earlier in this Foreword, Stuart (PS4NASA)

Comparison of the Daedalus Ziggurat with the Ziggurat at Ur by Richard Coombs.

has unequivocally argued that as1120pyramid20smallue2.jpg was made from the digital NASA source file 5564.jpg. I have counter-argued that as1120pyramid20smallue2.jpg was scanned from an analog original Apollo era print— a fact now confirmed by the sycophant. So this village idiot had now categorically confirmed that I was right all along and Dr. Stuart (PS4NASA) Robbins was categorically wrong.

But wait, wasn't I wrong about KK Samurai/Terry James not being the source of as1120pyramid20smallue2.jpg? Doesn't that make me equally in error?

As I established earlier in this Foreword, Stuart (PS4NASA) has unequivocally argued that as1120pyramid20smallue2.jpg was made from the digital NASA source file 5564.jpg. I have counter-argued that as1120pyramid20smallue2.jpg was scanned from an analog original Apollo era print— a fact now confirmed by the sycophant. So this village idiot had now categorically confirmed that I was right all along and Dr. Stuart (PS4NASA) Robbins was and is categorically wrong. With friends like these, Stuart (PS4NASA)…

So in one fell swoop, not only has Stuart's (PS4NASA) sycophant failed to prove that as1120pyramid20smallue2.jpg came from Terry James' website, he has by his own Plain Statement of Fact proven that it was not conjured up from NASA's tampered digital

A Meso-American connection confirmed?

Thanks to Frank, I think this confirms the suspicions of many of us over the years...Our Archaeologists can't figure out who built all of those fine structures on our planet and how! Well, take a good look... Is this a natural structure? Think about it! by kksamurai

image "5564.jpg," as Stuart Robbins (PS4NASA) has categorically declared.

Thanks for the help there buddy. But I bet you're off Stuart's (PS4NASA) Christmas card list now.

But more important is what Terry's version of the Ziggurat actually shows. First, in comparing as1120pyramid20smallue2. jpg to Terry's original work from his website in 1999, it is plain to see that all the same features appear in both. The left side and rear walls, the walled enclosure for the "temple," the square temple itself along with the entrance ramp and the dome on top. The left wall may be brighter and more defined in as1120pyramid20smallue2. jpg, but that may simply be because whoever enhanced it used a different technique than Terry James did. This is reinforced by the fact that the photographic glue residue overlays the top of the left wall, an impossibility if the wall had been "drawn-in" after scanning, as Stuart (PS4NASA) and the sycophants have argued (see enhancement above).

as1120pyramid20smallue2.jpg (L) and the version from Terry James website (R). Note that the "KK" watermark does not appear on "as1120pyramid20smallue2. jpg" and it is rotated and cropped differently.

Shortly after this exchange, Terry James himself showed up on the sycophant's website and called me and Mr. Hoagland out for not giving him credit for the Ziggurat. I later learned that he had apparently not actually seen the as1120pyramid20smallue2.jpg image when he made this declaration. Fortunately, he left his email on the blog post, so I wrote to him explaining the chain of custody of as1120pyramid20smallue2.jpg and why he hadn't been credited in *Ancient Aliens on the Moon*. I quickly got a reply, and we began a cordial email exchange. He made several major points in these emails:

"I can assure you I am neither a hoaxer or a Christian and yes I did find this pyramid on a very large scan of a photo sent to me by Frank Gault. I am also very certain that Frank did not mess with that scan or any of the many other scans he sent me... I also noticed that many of the lunar scans that Frank sent me were not found in the NASA public archive. I also noticed that some of the available images in the public archive had features removed or brushed out that were very clean in Frank's scans."

He also stated that he didn't always watermark his images, although all of the Ziggurat images on the archived site do have one. So as1120pyramid20smallue2.jpg could still have originated

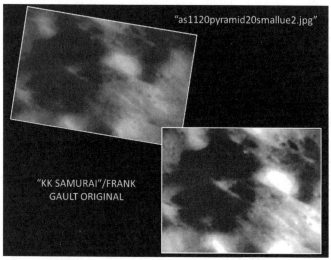

"as1120pyramid20smallue2.jpg" and Terry James' enhancement from the archived "KK Samurai" website. "as1120pyramid20smallue2.jpg" rotated to match alignment.

with him or Frank Gault:

"I should (also) point out that I didn't always put my watermark on my images. I often just signed them or didn't mark them at all. Sometimes I just inserted my name into the image file. And on occasion I would send someone a clip of my raw data provided they gave me credit for it. So it is probable that I sent out a clip of the pyramid to someone who renamed the file and used it. Also keep in mind that Frank may have also sent some of the data to someone else without my knowledge. After all he did have the scanned data on file."

Then, a second critic/attack dog that frequently collaborates with sycophant #1 sent an email claiming that Terry James had in fact admitted to "faking" the Ziggurat image. "Oh and isn't it great that Terry James aka KKsamurai has shown up. You know..., the guy who faked the ziggurat. He admitted he created it, still has the original, and has called out you and Mike as thieves and liars. Mike even dedicated almost an entire chapter to this fake." Knowing all of these statements to be false, I passed the email along to Terry, who quickly replied: "I've been there before. I did not fake that ziggurat. He's basing his position on a public archive image whereas I have real first generation data from a NASA lunar scientist. It's not worth arguing. I should also say I've found a lot of very interesting data on more interesting stuff... I can't help but say I have a lot more data yet unrevealed and none of it is faked. In fact most of it proves that NASA has withheld. Worse, they've edited many lunar images to hide the truth from the general public."

In other words, Terry James reinforced that not only are the public archives which Stuart (PS4NASA) and the sycophant's are so dependent on not complete, he also agreed with me that they have been altered from their original form when compared to first generation photographic prints. When I informed Terry James that I was going to cover all this in the Foreword to my new book, he kindly offered to send me his original scan sourced from Frank Gault, and also confirmed that as1120pyramid20smallue2.jpg is NOT an image from his site, as claimed by sycophant #1. "Mike. If you are putting this in a book you need data from the original scan... The original scan is darker. To lighten something for the sake of vision and perception is OK provided you give them the original data."

About a week later, I received a thumb-drive with Terry's original scan, a GIF file named "Apollo-AS11-38-5564.gif." The image has pixel dimensions of 1500 x 1138 at 72 DPI, making for an on disk file size of 1.62MB and a document size of 20.833 inches in width and a height of 15.806 inches. In other words, there's plenty of data well above the "limits of resolution" with which to determine the authenticity of the Ziggurat as an artificial structure. And that's exactly what it does.

In comparing as1120pyramid20smallue2.jpg with Frank Gault/Terry James' original scan, all the major features are again confirmed. The front wedge shaped buttresses, the left and rear walls, the entire walled enclosure, the square "temple" structure, the entrance ramp, the "windows" on the side, the dome on top— all of it. With a little enhancement work, it becomes even clearer.

What Terry's original scan shows is that all of my original speculations about the origins of as1120pyramid20smallue2.jpg are verified—it did in fact come from a first generation photographic print in the personal collection of a former NASA employee,

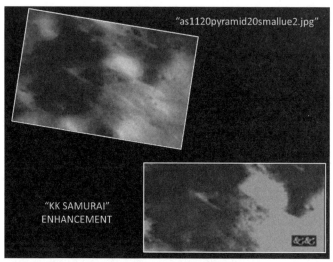

Ziggurat image "as1120pyramid20smallue2.jpg" (L) alongside Terry James' raw original (R).

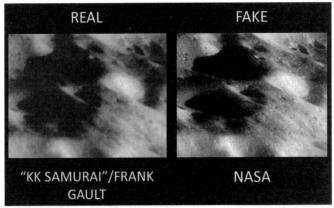

Comparison of Frank Gault/Terry James scan of the Ziggurat, and NASA
scan "5564.jpg,"— an obvious fake.

whether that person is Frank Gault's father or another source. It also
confirms that the current NASA version 5564.jpg, in total contrast
to the claims of Stuart (PS4NASA) Robbins, is an overt fake that
had the Ziggurat removed in a rather sloppy and obvious paint-over.
Given this, it is now safe to assume that ALL NASA digital imagery
is almost certainly compromised, as Terry has stated unequivocally
in his emails.

So, just to quickly recap:

1. There is not less noise in the NASA image 5564.jpg than
in the Ziggurat image as1120pyramid20smallue2.jpg. What
Stuart (PS4NASA) thinks is "noise" is actually photo-album
residue marks on the first-generation photographic print that
"as1120pyramid20smallue2.jpg" was scanned from.

2. Stuart's (PS4NASA) assumption that
as1120pyramid20smallue2.jpg was therefore modified after
NASA's "5564.jpg" and by his faulty reasoning manufactured from
it in Photoshop or a similar program is therefore falsified.

3. as1120pyramid20smallue2.jpg shows every indication of
being scanned from an early if not first-generation photographic
print, and therefore has an earlier derivation than 5564.jpg. This is
proven out by the research of Stuart's (PS4NASA) own fans.

4. Terry James aka "KK Samurai" has now produced an original
scan of a first-generation photographic print in the possession of Frank
Gault, the son of a former NASA lunar scientist who obtained the

31

photo directly from NASA. It shows that as1120pyramid20smallue2. jpg is far closer if not identical to the original NASA photograph AS11-38-5564. "5564.jpg" is therefore proven to be a fake digital image, at least as far as the Ziggurat is concerned.

5. Stuart's (PS4NASA) claim that "upsampling (interpolating) makes an image better is factually and demonstrably false" is shown to be factually and demonstrably false. Interpolation improves the quality of an image, as proven by the NASA images shown.

6. Terry James aka "KK Samurai" has categorically stated that the original Daedalus Ziggurat scan as1120pyramid20smallue2.jpg came from Frank Gault, and that neither he nor Gault altered or tampered with it in any way.

And all this pretty much blows point #1 out of the water.

Pillar #3 – Shadows and the Geometry of Light

In this section, I will address Stuart's (PS4NASA) claim that the Ziggurat image is a "fraud" because in his opinion, the lighting geometry is "wrong."

What Stuart (PS4NASA) said:

3. Why the shadowed parts of his ziggurat are lit up when they're in shadow, on top of a hill, and not facing anything

The Daedalus Ziggurat from NASA image "5564.jpg"

that should reflect light at them?

Response: I guess that we just fundamentally have a disagreement about this Stuart (PS4NASA). Again, he keeps repeating this claim, but no matter how many times he does, he can't make it true. They *are* facing "something" that would reflect light at them and account for the lighting geometry in as1120pyramid20smallue2.jpg. I see brightly lit hills all around the depression on the left side of the Ziggurat that would reflect light into the shadowed area under dispute.

In his original post, Stuart (PS4NASA) claimed that ALL of the shadows on the Moon are absolute black, and there are few (if any) gradations of light and dark on the Moon because of a lack of atmosphere. *"If you're in shadow, you're in shadow and it's going to be pitch-black (or almost pitch-black)."* He repeats this claim in a video he made attacking the Ziggurat as a hoax and calling Mr. Hoagland incompetent and a liar.[12]

As I pointed out in *Ancient Aliens on the Moon*, this is categorically untrue. I have seen hundreds, if not thousands, of lunar images where the shadows are far from "pitch-black (or almost pitch-black)." You actually have to be pretty dumb to even make such a claim. And when this inconvenient fact was pointed out to him on his YouTube video in the comments section, even Stuart (PS4NASA) agreed that this claim was, as he likes to put it, "factually incorrect."

You "thought of that an hour after he posted the video?" Really? So you *knew* that your statement that *"If you're in shadow, you're in shadow and it's going to be pitch-black (or almost pitch-black)"* was factually incorrect, but you just forgot to correct it?

I'm sorry Stuart (PS4NASA), did I miss something? I thought you were an astrophysics expert and all that stuff? I mean, wouldn't a Brainiac like you with all those important degrees who boasts about spending all that time processing images of the Moon have *known that from the get-go?* I did, but when I pointed it out in a blog post,[13] you asserted that I didn't know what I was talking about. And then, even after you "remembered" that your claim was false, you not

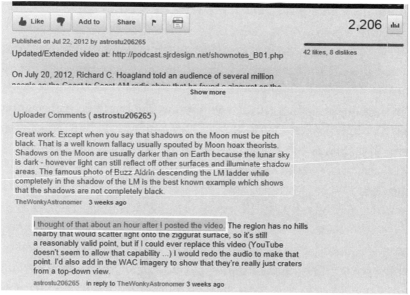

only didn't retract it, but you still made it a structural pillar of your claim that the Ziggurat image is "a hoax." You even posted another, longer video[14] later on still making the same false claim after you admit "remembering" there *is* such a thing as light scattering on the Moon.

So in other words, Stuart (PS4NASA), you *knew* from July 22nd 2012, when you "remembered" that light scatters into shadows on the Moon, that these claims were false. Yet, you never uploaded a video correcting this claim, and you still continue to defend the claim on your blog.

And you have the guts to impugn the integrity of myself and Mr. Hoagland? Look in a mirror pal…

Now, on to Mars…

[Editor's Note: This Foreword contains Mike Bara's full refutation of Dr. Robbin's point #1. To read the complete discussion and argument between Mike Bara and Dr. Stuart Robbins please go to Mike Bara's blog at:

MikeBara.blogspot.com

(Endnotes)

1 http://about.sjrdesign.net/index.html

2 http://www.colorado.edu/news/releases/2012/06/11/cu-boulder-research-ers-catalog-more-635000-martian-craters

3 http://pseudoastro.wordpress.com/2012/08/07/mike-bara-defends-the-lunar-ziggurat-my-response/

4 http://pseudoastro.wordpress.com/2012/08/16/kaguya-selene-%e3%81%8b%e3%81%90%e3%82%84-photographs-of-the-moon-specifically-the-claimed-ziggurat-area/

5 http://mikebara.blogspot.com/2012/08/the-daedalus-ziggurat-rubutting-dr.html

6 http://www.lpi.usra.edu/resources/apollo/frame/?AS11-38-5564

7 *Ancient Aliens on the Moon*, pp. 218

8 http://www.nasa.gov/images/content/673727main_PIA15980-full_full.jpg

9 http://www.uahirise.org/images/2012/details/cut/MSL_EDL_sharp.tif

10 http://www.unmannedspaceflight.com/index.php?showtopic=7397&st=330

11 http://web.archive.org/web/20001001005214/http://www.kksamurai.com/moon/pyramid1.html

12 http://www.youtube.com/watch?v=zzCxQLCNz4A&feature=youtu.be

CHAPTER 1
MARS AS THE ABODE OF LIFE

Mars as it once was, with a viable atmosphere and flowing, liquid oceans.

According to the mainstream astronomical view of Mars, it is nothing more than the lifeless, rocky and rather unremarkable 4th planet in our solar system which formed or "accreted" pretty much in its orbit as we see it today. Like Earth, it has an atmosphere, but Mars' atmosphere is much thinner (mean surface pressure is only about 0.6% that of Earth's) and 95% poisonous carbon dioxide. The diameter of Mars is 4,222 miles, about 53% of Earth's, which is 7,926 miles. Even though it is more than half Earth's size, Mars has only about 10% of Earth's mass, and Mars' gravity is only

37

38% that of Earth. Unlike Earth, which has one large Moon, it has two tiny "moons" which are in reality nothing more than captured asteroids. Even though Mars is absolutely classified as a planet by the astronomers, the truth is it is much closer to some of the larger Jovian and Saturnian moons in size, composition and density than it is to the rocky, dense inner planets Mercury, Venus and Earth. There is a reason for this, and we will cover that in due course.

Earth/Mars size comparison

Mars also has a number of unusual if not outright bizarre surface features which speak to a cataclysmic past. The Hellas Basin in the southern hemisphere is a massive impact depression that has caused some to theorize an object as large as a dwarf planet (like Pluto) may have created it. *Valles Marineris*, a huge gash running from east to west across the planetary sphere, is longer than the continental United States, 120 miles wide and 23,000 feet deep. *Olympus Mons*, standing atop the massive Tharsis bulge in the northern hemisphere, is almost three times taller than Mount Everest and at 14 miles high is the largest volcano in the solar system.

38

Topographic map of Mars showing the Hellas Basin (R) and Valles Marineris and Olympus Mons (L). The "line of dichotomy" runs the entire circumference of the planet.

Mars also has two distinctly different hemispheres, the heavily cratered southern and cue ball smooth northern. This separation is so dramatic that the boundary between them is referred to as the "line of dichotomy," a sharp feature with very little transition from one face to the other. It's almost as if Mars is two distinct planets; one even more cratered than the Moon, and the other smooth as a baby's bottom. The reasons for this are poorly understood by mainstream astronomers, and resultantly they have come up with several weak and contradictory explanations for how it could have happened. As we will see in the next chapter, there is a much better theory for Mars' two faced appearance that is as indisputable in its reality as it is unacceptable to the mainstream.

The traditional view of Mars' formation and its geologic history is one that we have all heard over and over, repeated ad nauseum by the various space and science cable channels that constantly regurgitate the same tired, shopworn copy. According to them, Mars formed out of an "accretion disk" 4.5 billion years ago, basically in the same orbit it is in now, and was bombarded by debris to the point that it became heavily cratered, like the Moon. Mars might have been warm and wet once, they say, and may even have supported some kind of simple life. But that was billions of years ago, and Mars' atmosphere was mostly swept away by some ancient, primordial impact event at least 4 billion years in the past, during the so-called "late heavy bombardment."[1] And, it has been a dead, cold world ever since.

But the reality of Mars is somewhat more interesting, if not incredibly more exciting.

The ancients knew as it one of the seven planets — or "wanderers" as they were known by the Greeks — visible without a telescope. These "Classical Planets," the Sun and Moon and the five non-earth planets of our solar system closest to the sun (Mercury, Venus, Mars, Jupiter, and Saturn) all have extensive mythical histories. Mars, because of its red color (caused by an iron-oxide rich surface layer of dust), was often associated with war and indeed "Mars" is the name for the Roman God of war. Its name may have been derived from the even earlier Sumerian God *Marduk*, who like the Roman Mars, was something of a god of war. In Egyptian mythology, the Sphinx God Horus was also frequently associated with the planet Mars, and in fact the Sphinx was at one time painted red in honor of this connection. They also shared a name in the ancient Egyptian tongue, "*Hor-Dshr*," literally "Horus the Red."

But Mars had been observed and tracked for centuries before Roman times.

As early as 1534 BC Egyptian astronomers had tracked the journey of the planet through the night sky to the point that they understood its apparent retrograde motion, a phenomenon where Mars actually seems to track backwards along its orbital path from the perspective of an Earth-based observer. Babylonian astronomers, between 626 and 539 BC, noted and recorded that Mars made 42 circuits of the zodiac every 79 years.[2] In the 4th century BC, the Greek astronomer/philosopher Aristotle observed Mars disappearing behind the disk of the Moon and concluded that it therefore must be farther away. Chinese astronomers were aware of Mars from at least the 4th century BC forward.

Modern observations of Mars really began in 1610, when Galileo

Galilei first observed the planet through a primitive telescope. Later, using a more sophisticated telescope, Dutch astronomer Christiaan Huygens became the first person to identify light and dark features on Mars, which he attributed to seasonal variations in water flows and vegetation. By the 19th century, the technological advancement of telescopes had reached a level good enough for surface features to be vaguely identified. But these kinds of observations could only be done once every two years or so.

Comparative orbits of Earth and Mars

Because of its highly "eccentric" or elliptical orbit (egg shaped as opposed to circular), Mars' distance relative to the Earth varies a great deal. In fact, Mars' orbit is so elliptical that its distance to the Earth can be as much as 249 million miles at its farthest to as little as about 34 million miles at its theoretical closest approach. The mechanics of this are somewhat complicated, but worth considering.

Because the Earth is closer to the Sun, its orbit is considerably shorter than Mars', about 365 days. Mars' orbit by comparison is about 687 Earth days, so a Martian "year" is almost twice as long.

Again, because Mars' orbit is dramatically more elliptical and slower than the Earth's, it accounts for the vast percentage of the space between the two worlds at any given moment. But Earth's orbit is a tiny bit elliptical too, and because of this it is sometimes closer to the Sun and sometimes farther away. This farthest point in the orbit is called "aphelion" by astronomers, and the nearest point in an orbit is called "perihelion." The Earth comes closest to the sun every year around January 3, and it is farthest from the sun every year around July 4. This equates to a distance of about 94.1 million miles at aphelion, and a distance of about 91.4 million miles at its closest approach, or perihelion. This means Earth's relative distance to the Sun varies by only 3.1 million miles in the course of one orbit (year).

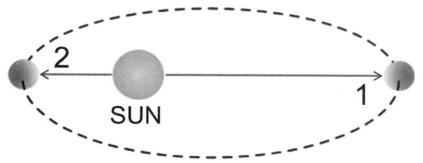

Typical planetary orbit showing aphelion (1) and perihelion (2)

By contrast, Mars' orbit around the Sun varies from 154.9 million miles at aphelion to 128.4 million miles at perihelion. This means that its relative distance to the Sun varies by some 26.5 million miles, compared to the Earth's mere 3.1 million miles. Obviously, Mars' orbit is far more eccentric by several orders of magnitude.

Earth (L) and Mars (R) at aphelion

The two worlds are therefore farthest apart (249 million miles) when they are on opposite sides of the Sun and both at their aphelion points. Conversely, they are at their closest to each other when they are on the same side of the Sun and both at their perihelion points, a distance which can be as little as about 34 million miles. Obviously, this would be the ideal time to make telescopic observations of Mars or launch probes to the Red Planet, so scientists gave these occurrences still another name, "perihelic oppositions." Oppositions occur about once every 780 days, a little over two years. Also, oppositions always occur within a few days of Mars' closest approach to Earth. But in general, closest approach and "opposition" are regarded as the same thing. In other words, Mars' closest approach in a given orbit always occurs around the time of an opposition.

It's important to note that not all oppositions of Mars are when it is near the perihelion point. Again, because of the eccentricity of Mars' orbit, the distance between the two worlds can vary widely at opposition. Between 1995 and 2007, there were seven oppositions of Mars, ranging from a distance of 63 million miles apart all the way down to 35 million miles in 2003. Of those oppositions, only the 2003 event was considered "perihelic." Indeed such events are

Illustration of Mars and the Earth at perihelion, or "opposition."

relatively rare. Only six occurred in the 20th century.[3]

One significant historic perihelic opposition of Mars took place on September 5, 1877, when Mars was only 35 million miles from the Earth. The discovery of Mars' two moons, Phobos and Deimos (which are really just captured asteroids) by Asaph Hall came during this opposition. This epic celestial event was also witnessed by a number of other prominent astronomers, among them Italian Giovanni Schiaparelli. Using an 8.7 inch telescope, Schiaparelli made observations that created a sensation when he released his hand drawn map of what he had seen.

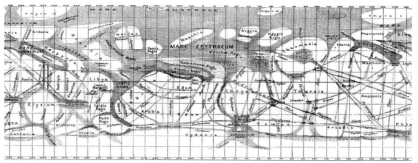

Hand drawn map of Mars by Giovanni Schiaparelli (1877).

Spotting numerous dark streaks which seemed to connect various major features of the planet, he described these streaks in his papers as "canali," an Italian word which can mean "canals." It is popularly stated in mainstream media that this was a mistranslation and that what he actually meant to describe were "channels" or "grooves" on the Martian surface. In fact, Schiaparelli never issued such a correction, and even after a worldwide firestorm erupted because of his claims, Schiaparelli still maintained that the translation as "canals" was correct. Or at least he never retracted it. As if to underscore this point, Schiaparelli even gave his "canali" names of famous rivers on Earth, indicating he did indeed see them as water transportation canals or aqueducts. In his 1893 book *Life on Mars*, Schiaparelli ended any question about his intent in using the word "canali.":

"Rather than true channels in a form familiar to us, we must imagine depressions in the soil that are not very deep, extended in a straight direction for thousands of miles, over a width of 100, 200 kilometers and maybe more. I have already pointed out that, in the absence of rain on Mars, these channels are probably the main mechanism by which the water (and with it organic life) can spread on the dry surface of the planet," he wrote.

Needless to say, this created a great deal of speculation as to just who these Martian canal builders might be. Popular magazines and newspapers quickly began to speculate about what life might be like on Mars. The idea that it might be inhabited by intelligent beings capable of planet-wide mega-engineering projects took hold and spread among the general public.

Other astronomers also observed the canals, foremost among them Percival Lowell. He built a massive observatory near Flagstaff, Arizona in 1894 which contained 11.8 and 17.7 inch telescopes which he used to observe the Red Planet for over 15 years. Like Schiaparelli, he too spotted the canals and created maps from his own observations.

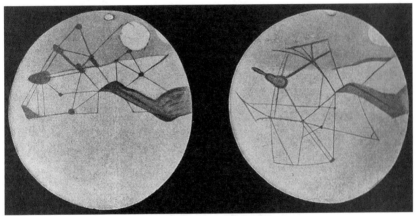

Two views of Mars as drawn by Percival Lowell.

Ultimately, Lowell wrote three books on Mars, of which *Mars as the Abode of Life* was probably the most influential. In these works, Lowell speculated that the canal system was used to drain melting water from the polar ice caps to various "oases," dark nodes that seemed to form at the intersections of the canals and varied with the seasons. This observation of a seasonal "wave of darkening" in the Martian summer may yet turn out to be the most important observation Lowell actually made (more on that later). In any event, Lowell's depiction of a desiccated, dying Mars and its desperate inhabitants trying to survive through planetary engineering on a massive scale captured the popular imagination. There is no question that Lowell and Schiaparelli's vision of Mars was the inspiration

45

for H.G. Wells' 1898 novel *War of the Worlds*, and indeed many other scientists and astronomers made efforts to confirm Lowell's work, with varying degrees of success. The observatory he built near Flagstaff was later used by Clyde Tombaugh to discover Pluto in 1930.

Mainstream science of course, as it always does, moved quickly to quash the idea of a habitable Mars and by the time Lowell died in 1916, the "canals" idea was under heavy fire. But in between Schiaparelli's discovery and Lowell's death, there were a few more

Tesla's lab and "wireless transmitter/receiver" on Long Island, NY.

notable and notably weird observations involving Mars.

In 1899, while working in his lab in Colorado Springs, Colorado, Serbian inventor and physicist Nikola Tesla was using one of his exotic Tesla Coil Receivers to track storms in the region. At one point, he began to hear regular, repetitive signals coming through his listening device. He publically speculated that these signals may have come from Mars, since it was in opposition to Earth at the time. Tesla further discussed the signals in the March, 1901 edition of *Collier's Weekly* magazine and the emotional effect it had on him:

> As I was improving my machines for the production of intense electrical actions, I was also perfecting the means for observing feeble efforts. One of the most interesting results, and also one of great practical importance, was the development of certain contrivances for indicating at a distance of many hundred miles an approaching storm, its direction, speed and distance traveled....
>
> It was in carrying on this work that for the first time I discovered those mysterious effects which have elicited such unusual interest. I had perfected the apparatus referred to so far that from my laboratory in the Colorado mountains I could feel the pulse of the globe, as it were, noting every electrical change that occurred within a radius of eleven hundred miles.
>
> I can never forget the first sensations I experienced when it dawned upon me that I had observed something possibly of incalculable consequences to mankind. I felt as though I were present at the birth of a new knowledge or the revelation of a great truth.... My first observations positively terrified me, as there was present in them something mysterious, not to say supernatural, and I was alone in my laboratory at night; but at that time the idea of these disturbances being intelligently controlled signals did not yet present itself to me.
>
> The changes I noted were taking place periodically and with such a clear suggestion of number and order that they were not traceable to any cause known to me. I was familiar, of course, with such electrical disturbances as are produced by the sun, Aurora Borealis, and earth currents, and I was as sure as I could be of any fact that these

variations were due to none of these causes. The nature of my experiments precluded the possibility of the changes being produced by atmospheric disturbances, as has been rashly asserted by some.

It was sometime afterward when the thought flashed upon my mind that the disturbances I had observed might be due to an intelligent control. Although I could not decipher their meaning, it was impossible for me to think of them as having been entirely accidental. The feeling is constantly growing on me that I had been the first to hear the greeting of one planet to another. A purpose was behind these electrical signals...[4]

In later years, Tesla was said to be developing a device called a "Teslascope" specifically for the purpose of communicating wirelessly with beings from other worlds, and especially Mars.

In a *New York Times* article, also published in in 1901, Edward Charles Pickering, then director of the Harvard College Observatory, seemed to come out in support of Tesla. He publicized a telegram the Observatory had received from the Lowell Observatory in Flagstaff that implied Martians were indeed trying to communicate with humans:

Early in December 1900, we received from Lowell Observatory in Arizona a telegram that a shaft of light had been seen to project from Mars (the Lowell observatory makes a specialty of Mars) lasting seventy minutes... The observer there is a careful, reliable man and there is no reason to doubt that the light existed. It was given as from a well-known geographical point on Mars. That was all... Whatever the light was, we have no means of knowing. Whether it had intelligence or not, no one can say. It is absolutely inexplicable.[5]

Pickering's shaft of blue light was never officially confirmed, but he obviously believed the report to be true. He even suggested creating an array of mirrors in the plains of Texas to reflect light towards Mars in a similar manner in hopes of establishing communications. Tesla, for his part, ridiculed Pickering's idea, insisting that there was a much more efficient method of talking to the Martians:

> The idea naturally presents itself that mirrors might be manufactured which will reflect sunlight in parallel beams. For the time being this is a task beyond human power, but no one can set a limit to the future achievement of man... But there is one method of putting ourselves in touch with other planets... This combination of apparatus is known as my wireless transmitter... It is evident, then, that in my experiments in 1899 and 1900, I have already produced disturbances on Mars incomparably more powerful than could be obtained by any light reflectors, however large.[6]

Tesla continued to staunchly defend his observations, insisting that the signals could not have come from one of the other planets or an earthbound radio source:

> To be sure, we do not have absolute proof that Mars is inhabited [...] Personally, I have my faith on the feeble planetary electrical disturbances which I discovered in the summer of 1899, and which according to my investigations, could not have originated from the Sun, the Moon, or Venus. Further study since has satisfied me they must have emanated from Mars. Others may scoff at this suggestion...[of] communicat[ing] with one of our heavenly neighbors, [such] as Mars... or treat it as a practical joke, but I have been in deep earnest about it ever since I made my first observations in Colorado Springs... At the time, there existed no wireless plant other than mine that could produce a disturbance perceptible in a radius of more than a few miles. Furthermore, the conditions under which I operated were ideal, and I was well trained for the work. The character of the disturbances recorded precluded the possibility of their being of terrestrial origin, and I also

eliminated the influence of the Sun, Moon, and Venus. As I then announced, the signals consisted in a regular repetition of numbers, and subsequent study convinced me that they must have emanated from Mars, the planet having just then been close to the earth.

Although the source of Tesla's signals has never been positively established, except in his (admittedly genius level) mind, the incident may have been a major inspiration for another, even more bizarre experiment that took place a generation later.

In 1919, Tesla's great rival and protégé Guglielmo "William" Marconi (co-inventor, with Tesla, of the radio) added to the mystery by insisting that he too had recorded signals from Mars on his yacht Electra, which was a kind of floating super laboratory. In a *New York Times* article dated September 2nd, 1921,[7] J.H.C. Macbeth, manager of Marconi's Marconi Wireless Telegraph Company, stated at a Rotary Club luncheon covered by the paper that Marconi was thoroughly positive he had made contact with Martians. Marconi's claims had been dismissed by critics as either atmospheric interference or misinterpreted local radio sources, but Macbeth pointed out that both Marconi and other experts had discounted those criticisms. Marconi showed that the signals were far too regular and repetitive to be naturally occurring atmospheric interference, and they were in far too great a wavelength to be produced by local, earthbound radio transmitters. At that time, terrestrial radio sources were only able to send signals in a wavelength of about 17,000 meters, while the wavelength of the signals he detected were in a range of 150,000 meters, obviously far beyond anything that could be produced by human hands at the time. The signal also repeatedly transmitted a reference to the letter "V," which may be an oblique reference to a hypothetical "Planet V" which we will discuss in the next chapter.

Macbeth concluded his remarks by suggesting that if we could find a way to boost our radio signals to match the "Martian wavelength" of 150,000 meters (an achievement he considered only a matter of time) then it might be possible to send signals to the Martians in International (Morse) code along with pictures to help establish a common means of communication. He proposed sending a series of dots and dashes forming the word "tree" or "Man" along with a picture of the related object, and repeating this over and over again until we received a response in the same code. The events of the next few years would call into question whether this might have actually been done as a secret military experiment.

As Mars approached opposition in 1924, the astronomical community prepared for the event with great excitement. The 1924 opposition was a perihelic opposition, and the two planets would be as close together as they had been since 1804 and as close as they would be for another 78 years. Not until the 2003 opposition would Mars be so near the Earth.

One of the planned experiments was to be run by a friend of Percival Lowell, Dr. David Peck Todd, professor emeritus of astronomy at Amherst College, and involved a device called the Jenkins Radio Camera. Todd convinced the skeptical Jenkins

Todd and Jenkins with the Jenkins Radio Camera printer.

to construct a "radio photo message continuous transmission machine" which would record all emanations from the Red Planet and theoretically convert them to a visual output, so any Martian signals could be visually recorded and hopefully decoded.

This experiment was taken so seriously that the United States Navy sent out a request for full radio silence world-wide every hour on the hour for five minutes, and for astronomers to use whatever equipment they had to listen in for signals from Mars. This formal request, sent by then Chief of U.S. Naval Operations Edward W. Eberle, ordered all U.S. Navy stations and astronomers to report their findings over a three-day period from 2400 hours on August 21st, 1924 to 2400 hours on August 24th, 1924.[8] Although the few private commercial radio transmitters could not be held to these orders, the initial 36-hour period was declared "National Radio Silence Day," and most (if not all) terrestrial radio stations cooperated with the request. A full description of how the test was conducted is described by blogger John Townsend and derived from other sources:[9]

> Heading the operation for the Army was Major General Charles Saltzman and Admiral Edward W. Eberle, Chief of Naval Operations. World renowned code expert William F. Friedman, then Chief of the Code Section in the office of the Chief Signal Officer of the Army was standing by to translate any messages that may have come from Mars. Friedman became chief technical consultant to the National Security Agency in 1952 and two years later became the special assistant to the director of the National Security Agency.
>
> At the US Naval Observatory a radio receiver was lifted 3000 meters above the ground in a dirigible tuned to a wavelength between 8 and 9 kilometers to record Martian signals during the silent periods. The U.S. Naval Observatory worked with Amherst College's newly developed "radio photo message continuous recording machine" or "Radio-Camera" invented by Francis Jenkins.

This machine produced flashes of light that were recorded on film whenever an incoming radio wave was detected. On August 21st the Jenkins Radio-Camera was turned on. A roll of sensitized paper about thirty feet long and about 6 inches wide, slowly moved past a point of light modulated by electrical radio signals fed from an antenna on Todd's dirigible aimed at Mars.

After 36 hours of recording, the film was developed and Francis Jenkins, the machine's inventor, told a press conference that the device had received signals. In addition to a fairly regular arrangement of dots and dashes, clusters of signals which seemed to appear every thirty minutes on the film also showed 'a repetition at intervals of about a half hour what appears to be a human face.'

These strange patterns were reported to the public five days later on August 28, 1924 in the *New York Times*...

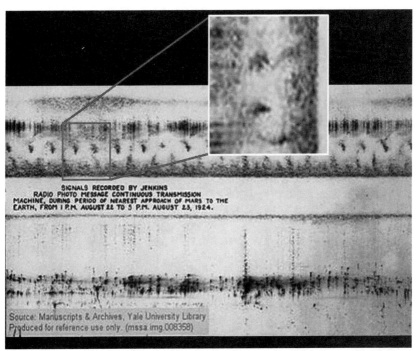

The signals received by the Jenkins Radio Camera in 1924. Inset: the "human face" repeating pattern rotated 90 degrees.

Now I don't know about you, but I find it completely fascinating that the Jenkins Radio Camera recorded *exactly* the kind of signal that Macbeth had described in his Rotary Club address several years before. Had the U.S. military taken note of Macbeth and Marconi's findings and secretly sent their own signal towards Mars during the 1922 opposition and then set this 1924 experiment up to see if the Martians answered back? If they did, they certainly seem to have gotten what they asked for.

The results were reported in the August 28th, 1924 edition of the *New York Times*:

> The development of the photographic film of the radio signals for the 29 hour period while Mars was close to Earth deepens the mystery of the dots and dashes heard by widely separated powerful stations. The film disclosed in black and white a regular arrangement of dots and dashes along one side. On the other, at about evenly spaced intervals, are curiously jumbled groups, each taking the form of a crudely drawn face.

Jenkins himself was at first reluctant to acknowledge the idea that the signals may have come from Mars, even though that was specifically what the experiment was set up to do. "I don't think the results have anything to do with Mars," Jenkins said. "Quite likely the sounds recorded are the result of heterodyning, or interference of radio signals." He could however, provide no proof for this claim, and was forced to admit there was no real explanation for the data received. "The film shows a repetition, at intervals of about a half hour, of what appears to be a man's face. It's a freak which we can't explain."[10]

Todd, however, was more decisive about what had been recorded: "The Jenkins machine is perhaps the hypothetical Martians' best chance of making themselves known to earth. If they have, as well they may, a machine that now is transmitting earthward their 'close-up' of faces, scenes, buildings, landscapes and what not, their sunlight values having been converted into electric values before projection earthward, all these would surely register on the weirdly unique little mechanism."

Unfortunately, the only record that seems to exist of the Jenkins Radio Camera output from the listening session is a crude scan done by the Yale University Library. Still, the dot-dash pattern and the creepy face are clearly visible, and just as eerie as you'd expect. The similarity to what Macbeth proposed in the *New York Times* as a means of establishing communication with alien life is striking. If Macbeth's idea to send a similar signal *towards* Mars was in fact done in secret during the 1922 opposition, then what Jenkins and Todd got back could be easily characterized as a "message received" response.

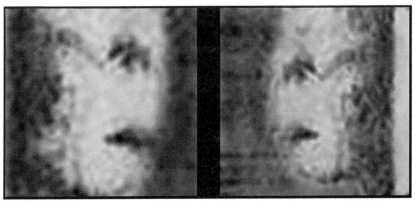

Close-up (enhanced) of facial profile image of "Voldemort" from the Jenkins Radio Camera printout.

But what bothers me the most about the printout is not that there does indeed appear to be the image of humanoid face in profile, it's with Jenkins' characterization of the face as "human." It in fact appears proto-human, with no real nose and some oddly shaped features on the skull around the eye socket. When I first saw the image, it struck me as familiar and it was only after several hours of studying it that I realized where I'd seen it before – in the Harry Potter movies.

It looked uncannily like English actor Ralph Fiennes as he was made up for the part of the arch villain Voldemort in the film adaptations of J.K Rowling's "Harry Potter" books. The more I

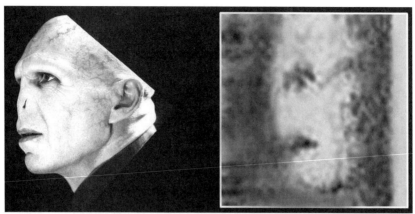

Two views of Mars as drawn by Percival Lowell.

looked, the closer the resemblance got and the more the hair stood up on my neck. Now I'm not for a minute suggesting that the Martians sent us a picture of Voldemort 80 years before the Harry Potter movies were made (although the usual village idiots will accuse me of it), but it does raise an interesting thought. Maybe Rowling or one of the filmmakers was aware of the Jenkins transmission and used the original Face on Mars as an inspiration. Or maybe, just maybe, this was a *real* transmission from Mars depicting the planets inhabitants as they once were. And maybe this transmission was an actual attempt at communication, either by a few desperate survivors or an automated beacon which had finally heard the SETI signal it had waited to hear for eons.

In either event, this first "Face on Mars" would turn out to be a pretty weird lead-in to what would come to light about the Red Planet decades later. But if the signal was real, and if it came from a machine that had survived over the centuries, it raised a serious question that still begs to be answered:

What happened to Mars?

(Endnotes)

1 http://en.wikipedia.org/wiki/Late_Heavy_Bombardment

2 Swerdlow, Noel M. (1998). *Periodicity and Variability of Synodic Phenomenon. The Babylonian Theory of the Planets*. Princeton University Press. pp. 34–72. ISBN 0-691-01196-6.

3 http://www.uapress.arizona.edu/onlinebks/MARS/APPENDS.HTM

4 Tesla, Nikola (February 19, 1901). "Talking with the Planets." *Collier's Weekly*. Retrieved on 2007-05-04

5 Professor Pickering (January 16, 1901). "The Light Flash From Mars." *The New York Times*. Archived from the original on 2007-05-20. Retrieved on 2007-05-20.

6 Tesla, Nikola (May 23, 1909). "How To Signal To Mars." *The New York Times*. Archived from the original on 2007-05-03. Retrieved on 2007-05-03.

7 http://query.nytimes.com/mem/archive-free/pdf?res=F60E14FB3F5D14738DDDAB0894D1405B818EF1D3

8 http://www.lettersofnote.com/2009/11/prepare-for-contact.html

9 http://themartianmystery.blogspot.com/2010/02/hidden-facts-radio-signals-from-mars.html

10 *Washington Post*, August 27, 1924. Weird "Radio Signal" Film Deepens Mystery of Mars Pictorially Recorded Messages Here Mere Tangles Mass of Dots and Dashes—Growing Wonderment May Bring Tenable Interpretation Theory.

CHAPTER 2
WHAT HAPPENED TO MARS?

MOLA reconstruction of the current surface of Mars.

In looking at the devastated surface of Mars, there is no question that something horrific and destructive overtook the planet in the distant past. This event (or series of events) was so cataclysmic that it stripped away most of the atmosphere, wrecked the magnetic field, left half the planet cratered and pock-marked and ruined any chance for higher life to survive there. Even mainstream astronomers agree that Mars was once warm and wet, and they even acknowledge the possibility that it once harbored simple microbial life. The only disagreements I have with them is just how advanced this life was and when this cataclysmic event took place.

As I will show you, Mars is not a planet but a moon—a former moon of a planet near its current orbit that doesn't exist anymore.

59

This moon we now know as Mars was warm, wet and hospitable, and would have been an ideal way-station for an alien species traveling to the equally hospitable Earth. But something happened to it, something that left Mars cold and desolate. The question is, what was that "something?"

As it turns out, that "something" is an idea so destabilizing to the mainstream way of thought that it has actually been deliberately suppressed for over a century—the complete destruction of at least two planets that used to exist in this solar system. Unfortunately for Mars and the highly advanced Ancient Alien civilization that once flourished on the Red Planet, they happened to be orbiting one of these planets at the time it exploded about 1.35 million years ago.

A lot of this can be better understood if we look at the current scientific theories for how Mars and the other planets and moons formed in the first place. According to these theories, planets, like our Earth, form from the dusty remnants of stellar nebulae created when a dying star explodes in a supernova. The "nebular hypothesis" (which has never actually been observed) argues that planets form in so-called "proto-planetary disks." Eventually tiny grains of dust from these stellar nebulae begin to clump together because of electrostatic charges in a process known as accretion. After a while, the accretion of the nebular material forms a clump about one mile in diameter. According to the theory, these clumps of material then begin to run into each other, somehow forming much larger clumps called planetesimals. This process then continues until the planetesimals get big enough to become classified as minor planets, like Pluto and Ceres. At that point however, the models begin to break down, because the number of collisions drops off considerably as the growing minor planets sweep the stellar nebula clean, like massive cosmic vacuum cleaners. This then requires a series of coincidental planetary collisions at thousands of miles per hour, which instead of breaking both objects into smaller pieces again, somehow magically manage to force and compress them into ever larger balls of material which then officially become planets. This supposedly explains why the inner, or "terrestrial" planets (Mercury, Venus, Earth and Mars) contain so much of the rare metals in the solar system. Since they formed in the innermost regions, in the so-called "habitable zone" where it was warmer, volatile molecules like water and methane couldn't condense. Only beyond 250 million miles out, in a region called the "frost line"—roughly the orbit of the asteroid belt—could these molecules become solid and eventually form planets like the gas giants of the outer solar system.

This is all very interesting, but wrong.

As I wrote about the accretion model in my second book *The Choice*:

> The biggest problem facing the accretion model— besides its never having been observed—is its inherent complexity. Not only does it require a series of magical planetary collisions as its base presumption, it also assumes, contrary to common sense, that these cataclysmic collisions (at thousands of miles per hour) would in the end be *constructive*, rather than *destructive*, to the nascent planets trying desperately to become big enough to take their place in the Solar System. This is hard to reconcile with actual observed results of this speculative process. Given that an equal amount of time has passed for all the newly forming bodies (and assuming a roughly equal distribution of stellar nebula material), they should all be about the same size. Only if some of the colliding bodies are substantially bigger than others is it possible for one planet to "eat" large chunks of another after a collision. How this would be possible given the size and composition limits imposed by the "frost line" idea is not made clear. Bodies of equal size, mass and speed would more likely obliterate each other completely. The accretion model also requires the planets to have highly eccentric (elliptical) orbits during their proto-planet phase, and this is hard to square with the observable planets' currently stable and decidedly un-eccentric orbits. Several hypotheses have been advanced to explain this problem away, including "gravitational drag" and "gravitational

wakes" created by smaller bodies, but no substantial experimental evidence exists supporting either idea.

So if the accretion model of planetary formation is falsified by this and other factors (see *The Choice*), then how did Mars and the other planets actually *get* here? And how can I prove that Mars is actually a moon of a missing planet, rather than a planet in its own right?

It's actually pretty easy. It's called the Solar Fission Theory of Planetary Formation. According to this theory, chiefly championed by the late Dr. Tom Van Flandern (Ph.D. Astronomy, Yale University), the planets were actually formed when the Sun spun them off from itself in the early days of the solar system. In his superb book *Dark Matter, Missing Planets and New Comets: Paradoxes Resolved, Origins Illuminated*,[1] Dr. Van Flandern argued persuasively for the Solar Fission theory.

According to the fission theory, a star forms when dust begins to condense into a single large clump in a proto-planetary disk and grows until it begins spinning, condenses and eventually ignites into a star. These young suns are spinning so fast that they eventually fling off large clumps of themselves as planets, which then spiral out from the star until they achieve stable orbits and begin to cool. Van Flandern states that the larger gas giant worlds would spin off first, in relative twin pairs, and generate multiple small rocky

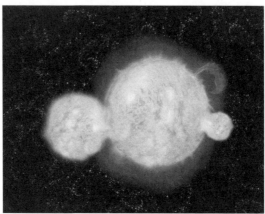

Artist's impression of the primordial Sun spinning off two proto-planets from its equatorial region (Krys Lilly).

moons. The last planets would be the inner rocky "terrestrial planets" like Venus and Earth, and they would spin off single large moons from their own molten planetary spheres.

The fission theory makes far more sense because it is consistent with most observations of our own solar system and others we have recently discovered. Only the fission theory can explain why all the planets are in the so-called "plane of the ecliptic," the equatorial plane of the Sun. If the accretion model was correct, planets would form all over the place and have orbits at all different angles to the Sun. It also explains why the planets have 98% of the spin energy in the solar system but only 0.002% of the mass. The Sun simply

Titius-Bode Law of Planetary Spacing		
Planet	Distance	Formula
Mercury	0.4	0.5
Venus	0.7	0.7
Earth	1.0	1.0
Mars	1.5	1.6
?	--	2.8
Jupiter	5.2	5.2
Saturn	9.5	10
Uranus	19.2	19.6
Neptune	30.1	38.8
Formula: distance in au $=0.4+0.3*2(n-2)$		

Chart showing Titus-Bode law of planetary spacing. Distances are in Astronomical Units.

"gave away" the vast majority of its spin energy to the planets in this birthing process, this being the price to pay for stability in energetic output (again, see *The Choice*). It would also explain why other solar systems have "Hot Jupiters," gas giant planets of Jupiter mass (or larger) that orbit their parent stars almost impossibly closely, sometimes even closer than the orbit of Mercury if they were in our own solar system. These "Hot Jupiters" alone are considered impossible by the standards of proof of the accretion model.

The fission theory also brings back another nearly forgotten scientific controversy from the past and places it in a new context. That context is called the "Exploded Planet Hypothesis," or EPH for short. The controversy is known as "Bode's Law."

Since it is covered extensively in *The Choice*, I will give the reader only a quick primer of the EPH here. The Exploded Planet Hypothesis re-ignites an age-old controversy from 1772 when an astronomer named Johann Elert Bode expanded on the earlier work of Johann Daniel Titius and showed that the planets' orbits should fit into a specific, resonant mathematical pattern. Upon comparing the projections with the actual distances of the planets then known, Bode found that all of them, Mercury, Venus, Earth, Mars, Jupiter and Saturn—fit the predicted pattern. The theory was given a huge boost in 1781, when Uranus was discovered exactly where Bode predicted it would be. The only problem was that according to Bode's Law, as it was now known, there should have been a fifth planet in an orbit where there was a gap between Mars and Jupiter. However, this gap also happened to coincide exactly with the location of the asteroid belt. In 1801, the dwarf planet Ceres was discovered in the belt, again exactly where Bode had predicted. It didn't take long after that for Heinrich Wilhelm Matthäus Olbers to suggest that the asteroid belt contained the exploded remnants of a missing planet, which he named "Phaeton."

Again, as it always does when a radical theory based on a catastrophic scenario appears in astronomy, mainstream science moved quickly against Bode and the critics seemed vindicated in 1846 when Neptune was discovered and deviated from the formula by about 29%. However, the issue was later complicated by the

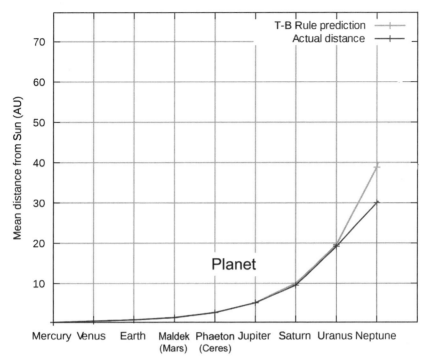

Mercury Venus   Earth   Maldek Phaeton Jupiter Saturn Uranus Neptune
(Mars)   (Ceres)

discovery of Pluto, which was found in a location almost exactly where Bode's Law would have predicted an eigth planet. Even with Pluto and Neptune included, the average margin of error according to Bode's formula is only about 14%, and Jupiter and the asteroid belt (Ceres) are virtually dead on.

Van Flandern explains these deviations away by speculating that Pluto is an escaped moon of Neptune, flung away when some ancient encounter (Nibiru anyone?) pushed Neptune out of its original resonant orbit and into its current, closer location. Such encounters were certainly far more common in the early days of the solar system, so this is hardly far-fetched. He also asserted that Mercury was not a planet, but rather an escaped moon of Venus, and that Mars was a remnant moon of a second exploded planet, which he called "Planet V" but which I dubbed "Maldek" in *The Choice*.

Even though Bode's theory had proven to be 88.88% correct, it was quickly dismissed as mere coincidence and several astronomers lobbied the (worthless) peer-reviewed journals to refuse to even review any papers discussing the theory. This held until very recently, when a paper by a group of astronomers at the European Southern Observatory from the University of Geneva was published in the August 13, 2010 issue of the international journal *Astronomy & Astrophysics*.[2]

Regardless, Van Flandern continued his work and discovered that long period comets (comets which have never entered the inner solar system before), long thought to be rouges that pass through the solar system on random trajectories, did not have random trajectories at all. Instead, the vast majority of them traced their orbits back to a single origin point. That origin point turned out not only to lie within the solar system, but in fact it was in the asteroid belt itself. The conclusion was inescapable. These comets had their origins from a single, cataclysmic event; the explosion of a former planet (Phaeton) that used to occupy the orbit of the asteroid belt.

Critics of course argued that was little to no evidence to support that Mercury and Mars were former moons, and in any event there was no mechanism to explain how a planet could explode. Debunkers and critics always make the "mechanism" argument when they can't really argue with the original observations, but

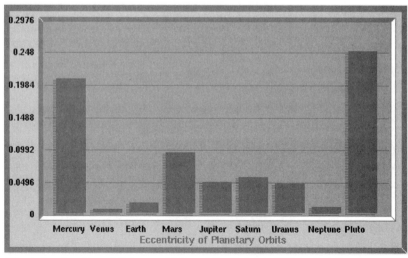

Chart showing the orbital eccentricity of the "planets." Note that the three most elliptical orbits belong to the three worlds suspected of being former moons, Mercury, Mars and Pluto. The implication is that they were pushed out of their more circular orbits by catastrophic events taking place within the solar system. (source:http://csep10.phys.utk.edu/astr161/lect/solarsys/revolution.html)

in any case Van Flandern has long since provided three different possible explosion mechanisms, all supported by the often-cited peer-reviewed papers: phase changes, natural fission reactions and gravitational heat energy. Van Flandern also pointed out that in *Dark Matter, Missing Planets and New Comets...*, there are more than 100 separate lines of evidence listed in support of the EPH and only two that seem, on the surface anyway, to contradict it.

The other major factor in support of the idea that Mars, Mercury and Pluto are recently escaped moons is their orbital eccentricity. These three bodies have by far the most elliptical orbits of any major bodies in the solar system. If they had been in their current locations for billions of years as the standard models demand, their orbit would have long since circularized.

Unfortunately, Van Flandern passed away before he could see his theories proven out. But that has not stopped the accumulation of evidence in favor of them. According to the EPH, planets should form in roughly twin pairs, which would explain the asteroid belt (debris from Phaeton, as Olbers suspected) and why for instance Mars' orbit is so eccentric. If in fact Mars was a moon of Maldek, which was destroyed in some unimaginable cataclysm, it would have then been pushed out of that original orbit and into its current (very) elliptical one. But it wasn't until fairly recently that astronomers came up with a theory that could explain just why Planet V and Planet K, or Maldek and Phaeton, as I've named them, are no longer

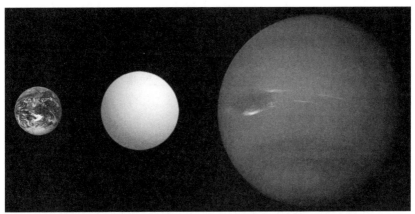

Size comparison between Earth, a "Super Earth," and Neptune.

with us.

It's because they were "Super Earths."

"Super Earths" are large, rocky terrestrial planets about twice the size and about five times as massive as our own Earth (at the low end), but with far more tectonic activity (earthquakes) and far less crustal stability. This makes them ideal candidates for at least two of the established "explosion mechanisms"—phase changes and natural fission reactors.

Super Earths are thought likely to have atmospheres and liquid water, providing they reside in the so-called "habitable zone" of their particular solar system, a region where they get enough heat from their parent star for liquid water to flow. In our own solar system, the "habitable zone" extends from the orbit of Venus all the way out to the asteroid belt, meaning that both Maldek and Phaeton could have possibly had flowing liquid water and possibly supported life.

There is a grand total (as of this writing) of 38 known Super Earths and Super Earth candidates identified by NASA's Kepler Mission.[3] So far, 16 of them have been found in the habitable zones of other star systems, meaning that 42% of all the observed Super Earths are in their respective solar systems' habitable zones. So there is every reason to think that it is possible that Super Earths once existed in our own solar system as well, and may have been able to support life. Certainly, as a moon of one of these Super Earths, Mars was well within the habitable zone itself and there is every reason to think that it could and did support life. These Kepler studies also fit right in with Van Flandern's model of where Maldek and Phaeton once resided (the orbit of Mars and the orbit of the asteroid belt, respectively).

The original configuration of the solar system

What this means is that the original solar system actually had at least eight planets: Venus, Earth, Maldek, Phaeton, Jupiter, Saturn, Uranus and Neptune, and several large moons, at least three of which are or once were mistaken for planets. In this scenario, Pluto is an escaped moon of Neptune, Mercury an escaped moon of Venus, Ceres a remnant moon of Phaeton, and Mars a remnant moon of Maldek. To the usual village idiots and mainstream astronomers that are funded by NASA, all this will of course have the air of heresy. But fear not, for I can *prove* it...

THE MARS TIDAL MODEL

The proof I speak of lies on Mars. In fact, it's written all over the face of it. The truth is, Mars was at one time warm, wet and hospitable, just as Schiaparelli and Lowell once dreamed. As a moon of Maldek, Mars enjoyed a comfortable location in the habitable zone of our solar system, probably for billions of years and up until very recently, in astronomical terms. Most likely, it was a very active world geologically because of the immense gravitational forces exerted on it by Maldek. It was also, as we will see in the coming chapters, the home of a vast and highly advanced civilization that was probably wiped out in a single day.

The standard model requires that whatever happened to Mars *must* have taken place billions of years ago. It assumes that the craters in the southern hemisphere below the line of dichotomy are from the theorized "late heavy bombardment" phase of the solar system at least 3.8 billion years ago. But the EPH doesn't require a "late heavy bombardment," and it says that Mars could have been Earth-like within only few million years ago, if that.

While the EPH goes very far in explaining many of the contradictions of the standard model of planetary formation and replacing them with something more in line with the actual state of the solar system, changing Mars' status from that of a full planet down to a mere moon of Maldek is something far more heretical. Merely raising the possibility elicits a reaction from mainstream astronomers akin to what you get when you show a crucifix to a vampire. That doesn't make it any less true, or any less provable.

Van Flandern himself generated a list of supporting observations from his own research which serves as an excellent starting point to a discussion of the issue:

- Mars is much less massive than any planet not itself suspected of being a former moon

- Orbit of Mars is more elliptical than for any larger-mass planet

- Spin is slower than larger planets, except where a massive moon has intervened

- Large offset of center of figure from center of mass

- Shape not in equilibrium with spin

- Southern hemisphere is saturated with craters, the northern has sparse cratering

- The "crustal dichotomy" boundary is nearly a great circle

- North hemisphere has a smooth, 1-km-thick crust; south crust is over 20-km thick

- Crustal thickness in south decreases gradually toward hemisphere edges

- Lobate scarps occur near hemisphere divide, compressed perpendicular to boundary

- Huge volcanoes arose where uplift pressure from mass redistribution is maximal

- A sudden geographic pole shift of order 90° occurred

- Much of the original atmosphere has been lost

- A sudden, massive flood with no obvious source occurred

- Xe129, a fission product of massive explosions, has an excess abundance on Mars

This is an impressive list of supporting data, and all of it fits with idea that Mars was once a moon of much larger, more massive planet, which I have named Maldek. But it doesn't prove it. Only one thing can do that...

Tidal bulges.

Tidal bulges on planetary spheres are something that every planetary astronomer is (or should be) familiar with. Earth itself has a tidal bulge caused by the Moon's gravitation tugging on it as it rotates around our planet in its roughly 27-day cycle. This is of course what causes (mostly) the high and low ocean tides. Other planets' gravitational fields cause similar tidal bulges on their moons, but in the case of the Earth-Moon system, the two worlds are tidally locked, meaning that the Moon always shows the same face (more or less, there's a little bit of libration) to the Earth at all times. That's why there's a "dark" or farside of the Moon that we

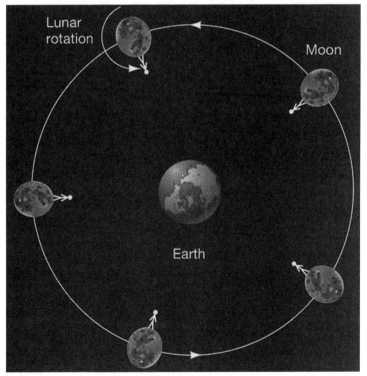

Illustration of tidal locked Moon as it orbits the Earth. Note (exaggerated) egg-shaped distortion of the Moon caused by Earth's gravity.

never saw until the 1960s when we sent space probes to photograph the pyramids, ziggurats and such which reside there.

There are similar tidal-locked arrangements all around the solar system. Jupiter has eight moons which always show the same face to their parent planet. Saturn has 15, Uranus five, Neptune two, and so on. But what is most important about each of these arrangements is that without exception, they all share one distinct, unique and undeniable geologic feature that defines their commitment to the tidal lock arrangement like a brand or a tattoo: tidal bulges.

Because of gravity and the various orbital mechanics involved, the one indisputable and inviolable characteristic that 100% of all tidally locked bodies share is bulges on their otherwise spherical surfaces, approximately 180 degrees apart. It is also well-understood that this indisputable signature of a tidal locked orbital condition is further characterized by the fact that the side facing the parent planet or moon, be it Jupiter, Saturn, Maldek or any other massive body, will have the *larger* bulge, and the opposite side, 180 degrees around the circumference of the sphere, the so-called "antipode" location, will have the *smaller* bulge. So the one way to recognize that a planet or moon is or once was in a tidal-locked relationship

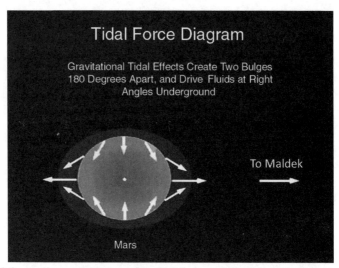

Tidal force diagram. Larger of 2 tidal bulges will always face the parent planet in a tidally locked relationship. A 2nd, smaller bulge will always appear 180 degrees around the circumference of the planet or moon.

with a more massive parent planet is two tidal bulges, one larger, the other smaller, 180 degrees away from each other.

Now, I cannot emphasize these points enough. 100% of the time, in absolutely every circumstance involving a tidal locked relationship between two celestial bodies, the tidal-locked body will have two tidal bulges at 180 degrees apart, and one will be larger than the other. It's a stone-cold, lead pipe lock.

Guaranteed.

Indisputable.

Inviolable.

Undebatable.

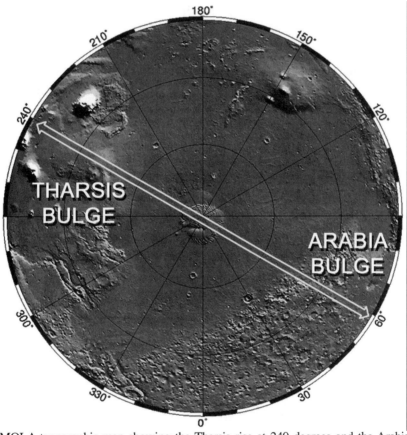

MOLA topographic map showing the Tharsis rise at 240 degrees and the Arabia bulge at 60 degrees, placing the two highest points on the Martian surface at antipodal locations—an indisputable signature that Mars was once in a tidal locked relationship with a parent planet (Maldek).

73

It is, as the saying goes, absolute proof beyond a reasonable doubt.

And Mars has them.

The *Mars Global Surveyor* spacecraft (MGS—1998-2006) had a device called the MOLA, or Mars Orbiter Laser Altimeter, that successfully mapped most of the surface of Mars. Using a laser (obviously) it was able to determine the highest and lowest points on the Martian surface and allow researchers to create a topographical map of the planet. The results were striking and consistent with the EPH's notion that Mars was once a former moon of Maldek. It turned out that the two highest points on the surface of Mars were a pair of huge mantle uplifts: the Tharsis bulge, which is home to Olympus Mons, and the Arabia Bulge, which is *exactly*... wait for

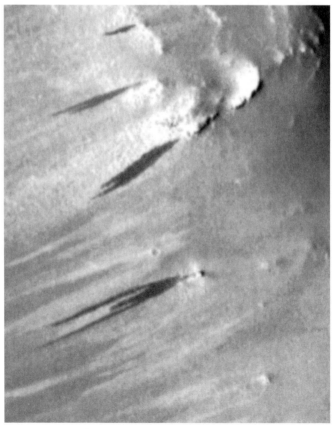

Water streaks on Mars (M04-00072).

it... *180 degrees around the circumference of Mars*. Beyond that, just as the tidal model predicts, Tharsis was much larger than the Arabia bulge, meaning there is no doubt they were both created by enormous tidal forces.

In other words, these two massive mantle uplifts were at antipodal locations on the Martian sphere—exactly where they should be if Mars was once in a tidal locked relationship with a larger, more massive body. Or to put it more exactly: Mars was a moon of Maldek, period.

And no, there is no other standard geological explanation that could account for this phenomenon. None.

This observation, first made by my co-author on *Dark Mission*, Richard C. Hoagland, came after he sent me an MGS image of a dark streak on a Mars. I immediately recognized the dark streak for what it was; it was a stream of water from a melting ice pocket on the surface. Intrigued, Hoagland looked into the dark streaks and found some amateur researchers had been looking at them too. He soon had two of the brightest, Efrain Palermo and Jill England, mapping the locations of images where they found the dark water streaks.

3D map showing the 180-degree relationship between stain images clustering preferentially on the Tharsis and Arabia bulges. Graphic by Efrain Palermo.

Immediately, a striking global pattern emerged: the water flows seemed to cluster preferentially around two pronounced geological features on the Martian surface, the Tharsis and Arabia mantle uplifts, or bulges. The curious thing about these bulges, though, was their location: 180° apart.

It was Hoagland who first realized the significance of this distribution. The scars of Mars' former life as a tidal-locked companion of a mysterious, long-forgotten parent planet (Maldek) told the story. Mars was not a planet. It was a moon. A moon that had once been in a tidally locked relationship with her parent planet Maldek, just as the Moon is with the Earth. And, it had been a moon that once had vast, liquid water oceans. And *that* meant one more thing was a virtual certainty—Life!

Immediately, all sorts of implications fell out of this inevitable conclusion. In our model, this tidal lock relationship went on for hundreds of millions of years, if not billions, and was broken only when Maldek was destroyed in (most likely) a gargantuan internal explosion.

MOLA topographic map of Mars showing the heavily cratered southern hemisphere below the "line of dichotomy."

The resultant debris bombarded not only Mars, but also a large portion of the solar system. Mars, as a close-by satellite, was the hardest hit, as the devastating impacts ripped away most of the planet's atmosphere and blasted it with rubble and debris at hypersonic speeds. It is this bombardment that accounts for the well-known "crustal dichotomy" of Mars, where the southern hemisphere has a crustal thickness nearly twice that of the northern lowlands in some places.

The standard model of Mars says that the heavily cratered region below the line of dichotomy is the oldest, created by the mythical "late heavy bombardment." It imagines that somehow the northern hemisphere had its surface ripped away and scoured clean by an as yet poorly understood event. Under this scenario, advocated by Graham Hancock and Robert Bauval in their book *The Mars Mystery*, a large object they named *Aster* struck Mars in the Hellas Basin, tore through the innards of the planet and created Valles Marineris as a huge tear, then blasted off the crust in the northern hemisphere along with most of the atmosphere. It's a possible scenario, but it doesn't explain the tidal bulges or a myriad of other geologic anomalies on Mars (see the Mars Tidal Model paper, available online).[4]

The tidal model not only explains more of the strange geologic factoids about Mars than the Aster scenario does, it also provides a plausible planetary environment for a civilization to have evolved in. In the tidal model, the water stains are simply fossil remnants of a once vast ocean that existed on the Red Planet. Held in place by the massive gravity of Maldek, Valles Marineris is explained as a massive tidal bore between the two bimodal oceans, scouring away the crust as it ran back and forth from Tharsis to Arabia. This means that the cratered area below the line of dichotomy was caused by debris from the explosion of Maldek, and is in fact the newer surface of the planet, not the older. The smooth-planed northern hemisphere above the line of dichotomy was created when this massive global ocean was released from the gravitational tug of Maldek and scoured the surface of the planet facing away from the impact zone. It later froze there, creating the smooth, nearly un-cratered surface we see today.

And killing Mars.

This catastrophe not only destroyed whatever civilization had once flourished on Mars, leaving only ruins, it pushed Mars into its current very elliptical orbit. While Mars currently has a very dramatic orbital eccentricity of 0.093 (more than five times that of Earth), it is known that in the very near past, only 1.35 million Earth years ago, its orbit was far more circular with an eccentricity of only 0.002, much less than Earth's today.[5] This is yet another line of evidence that shows that something very bad happened to Mars, and very quickly, as it turns out.

Since that horrific day, Mars has stayed in a state of near stasis, with its cool temperatures turning it into a kind of giant refrigerator and preserving whatever was left unscathed by the catastrophe Mars went through. The mainstream scientists will tell you that there isn't much of anything there, no buildings, no canals, not even any microbes. They will tell you that Mars has been dead for billions of years and that there is no evidence there was ever an advanced civilization there, whether indigenous or transient. What I will tell you is that there is an unbelievable amount of Ancient Alien evidence on Mars, that it was once covered by a vast and complex civilization, one that was far more advanced than ours. I will tell you despite what happened to it, Mars still lives. Life will always find a way. In the case of Mars, it certainly has. Microbial and even advanced plant life is a given. What we are about to see beyond that is evidence of something far more, something more akin to the Krell civilization of the classic Sci-Fi movie *Forbidden Planet*.

So let's go find it...

(Endnotes)

1   North Atlantic Books; Rev Sub edition (January 8, 1999), ISBN-13: 978-1556432682

2   *Astronomy & Astrophysics* manuscript no. HD10180, August 13, 2010

3   http://www.solstation.com/planets/super-earths.htm

4   http://www.enterprisemission.com/tide.htm

5   http://main.chemistry.unina.it/~alvitagl/solex/MarsDist.html

CHAPTER 3
EXPLORATION BEGINS

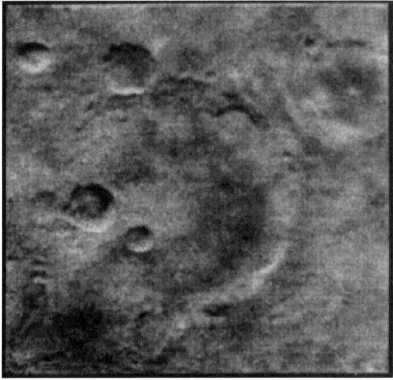

One of the first photographs of the surface of Mars, taken by *Mariner 4* in 1965.

By the dawn of the space age in the late 1950s, the prevailing scientific perspective on Mars was that it was a cold, dead and uninhabitable world. The tidal model and the notion of a recently warm, wet and hospitable Mars were still decades away. While a few scientists still held onto the "Lowellian" view that conditions might be harsh but still able to support higher life, that idea was pretty much quashed in 1965 when an early NASA probe named *Mariner 4* achieved the first flyby of the Red Planet and took the very first close-up views of another world. Because of the complications involved in spacecraft navigation at the time (see *The Choice*), *Mariner 4* was only able to take 22 pictures as it flew by Mars on July 14th and 15th, 1965. To the delight of the accretion model theorists, what *Mariner 4* photographed was a heavily cratered, Moon-like surface that they believed fit their models and led to the false assumption that Mars had been dead for billions of years—at least since the mythical "late heavy bombardment." At that time, they had no concept of Mars' crustal dichotomy or any real sense of its complicated and contradictory geologic history.

Other scientific instruments aboard the spacecraft were just as disappointing to the Lowell model and its advocates in the scientific community. *Mariner 4* found that Mars' mean surface atmospheric pressure was only 0.6% that of Earth, making it basically impossible (they thought) for liquid water to flow there. This finding killed the idea of Schiaparelli's canals, and any hopes for finding higher forms of life living above the surface were dashed when it was discovered that Mars' magnetic field was also basically non-existent, exposing any life to potentially deadly cosmic radiation. Further tests led to estimated daytime temperatures of -100 degrees Celsius.

Mariners 6 and *7* were launched toward Mars in 1969, and the twin flyby missions returned hundreds of photographs that allowed researchers to map about 20% of Mars' surface. By a quirk of their trajectories, once again all the photos were of the heavily cratered surface below the line of dichotomy, so planetary scientists still had no idea that Mars had essentially two different surfaces. It wasn't until *Mariner 9* achieved orbit around Mars in 1971 that humans got their first close up look at the complete surface in reasonably high detail. In 349 days in orbit, *Mariner 9* transmitted 7,329 images, covering a full 100% of Mars' surface area. This enabled NASA scientists to assemble the first photo mosaics of the Red Planet and led to the discovery of features like Valles Marineris and Olympus Mons.

THE PYRAMIDS OF ELYSIUM

Mariner 9 also led to one of the more interesting discoveries about Mars and led to the first modern speculation about the

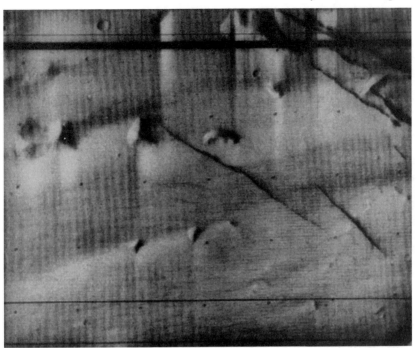

Carl Sagan's original "Pyramids of Elysium" presented in his book *Cosmos* in 1980.

possibility of an Ancient Alien presence. In looking over the images, the Patron Saint of Debunkery — none other than Cornell astronomer Carl Sagan himself — spotted some tetrahedral and polyhedral "mounds" in the Elysium Planitia region of Mars. He made quite a big deal out of what he eloquently dubbed the "Pyramids of Elysium" in both his 1980 book *Cosmos* and his PBS TV series of the same name. "The largest are 3 kilometers across at the base, and 1 kilometer high — much larger than the pyramids of Sumer, Egypt or Mexico on Earth," he wrote. "They seem eroded and ancient, and are, perhaps, only small mountains, sandblasted for ages. But they warrant, I think, a careful look."

Well yeah.

For starters, the Elysium location made a lot of sense as a post-apocalyptic site for an advanced, "Type II" civilization to rebuild.[1] Since it was on the lower, smoother northern plains of Mars, it was a relatively flat area where pyramidal structures could more easily be erected. Such massive structures could serve as ideal "arcologies," self-contained "architectural ecologies" along the lines proposed by

The Shimizu Mega-City Pyramid, a proposed arcology in China.

architect Paolo Soleri, capable of providing shelter and living space for thousands if not tens of thousands of beings amidst the newly harsh climate of Mars.

Such living structures would ideally be shaped like pyramids, either three-sided (tetrahedral) or four-sided (along the lines of the great pyramids of Egypt or Mexico). In looking at the Elysium region of Mars, it quickly becomes evident that the area is littered with

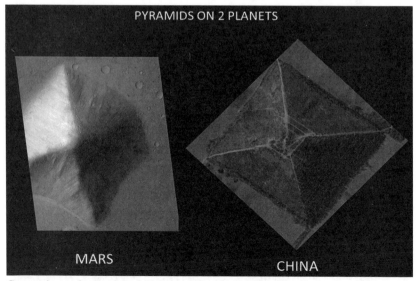

PYRAMIDS ON 2 PLANETS

MARS

CHINA

Comparison of one of Carl Sagan's "Pyramids of Elysium" on Mars (L) with an identical artificial structure in China (R). NASA image ESP_017574_1965.

such stand-alone three and four sided pyramidal structures. Some of the Elysium pyramids imaged at high resolutions bear a strong resemblance to Mexican and more recently discovered Chinese pyramids to the northwest of Xi'an, on the Qin Chuan Plains in Shaanxi Province. In fact, in at least one case, the resemblance is positively eerie.

Thought to be the tomb of Chinese Emperor Xuan of Han (91 BC – 49 BC), this pyramid is estimated to be somewhere between 200-400 feet high (measurements are hard to come by as the secretive Chinese government hasn't allowed much in the way of excavation of their ancient monuments). This obviously pales in comparison to Sagan's description of his "Pyramids of Elysium" being between one and three kilometers high. But what's important to note is that while there is a difference in *scale*, there is no difference in *form*.

While the pyramid in China is indisputably artificial, its

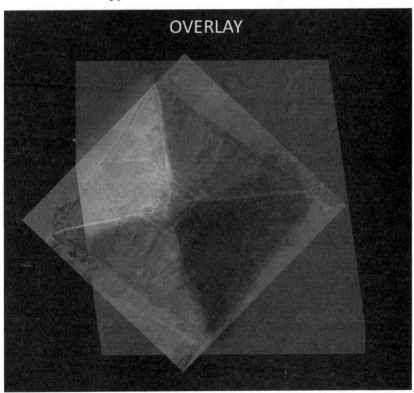

Overlay of NASA MRO image ESP_017574_1965 and Chinese pyramid located in the Shaanxi Province in China.

resemblance to the pyramid imaged in the Elysium region by NASA's *Mars Reconnaissance Orbiter* (MRO) in April, 2010 is equally indisputable. NASA is unequivocal in their stance that the Elysium pyramid is a natural formation. But in fact it is virtually identical in form to the tomb of Emperor Xuan. So how can NASA be so certain of the "natural" origins of it?

This becomes even more obvious when you simply overlay the two structures. While one of the corners is cut off the MRO image, it is quickly clear that both the structure and the sagging/ collapse pattern of these two ancient constructions are identical. This means that the Elysium Pyramid, which NASA flatly argues is a natural object formed by wind (a ridiculous impossibility which we will address presently), is eroding along *exactly* the same lines as its indisputably artificial twin pyramid in China.

Which is a pretty neat trick for a "natural" object.

But if the Chinese pyramids are confirmed as artificial, why can't Sagan's pyramids be artificial too, as he speculated? Well, they can be. And they *are*...

The weakest part of NASA's arguments about this and other pyramidal structures on Mars that we will study in future chapters is their complete isolation in the surrounding terrain. The Chinese pyramids, the Mexican pyramids and the Egyptian pyramids always stand alone, and clearly are not formed by wind, but by deliberate construction. All pyramidal mountains I've ever seen on Earth are part of mountain chains, formed by the upward surge of high pressure molten magma, like the recently discovered pyramidal mountains in Bosnia-Herzegovina. The fully isolated pyramidal forms are always artificial, at least on Earth. Unless there are some special rules of geology that apply only to Mars, how is it that Dr. Sagan can find tetrahedral and four-sided pyramids on Mars in complete isolation, and yet their origin must somehow be natural? Throwing aside the question of wind, which we will get to in the next section, if anyone can find me a verified natural three, four or five sided pyramid anywhere on Earth, completely isolated—well let's just say I'm open to it. But not in the 15 plus years I've been studying Mars has anyone claiming that these pyramidal structures are natural ever produced such an image.

All of this argument and speculation was to take an even more intense and interesting turn a few years later, when NASA launched the *Viking* series of probes to Mars.

THE "INCA CITY"

Another very interesting discovery by *Mariner 9* was a very strange, geometric formation near the South Polar Region. Informally

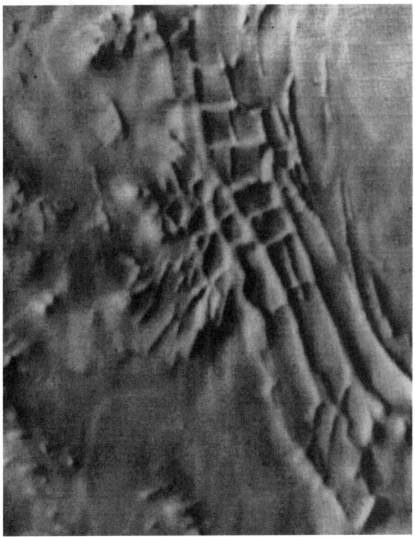

The "Inca City" as viewed by *Mariner 9* in 1972. NASA Frame Number 4212-15.

dubbed the "Inca City" by *Mariner 9* scientists in 1972, the Inca City was a set of intersecting ridges, uniform in height and at right angles to each other. Their origin has never been understood, though most NASA scientists tried (and failed) to attribute them to various natural processes. Sand dunes, dikes formed by injections of molten rock, or "soft sediments" injected into subsurface cracks that subsequently hardened and then were exposed at the surface by wind erosion were all proffered as possible explanations. NASA explained the geologic theories this way in 1998: "Possible interpretations that have been offered to explain the ridges include the following: (a) ancient sand dunes that became buried by polar ice and dust deposits and were lithified (cemented to make something like sandstone), (b) clastic dikes formed of sand and gravel that infiltrated down into cracks in

Partially buried ruins near Ashur, Iraq compare favorably to the Inca City on Mars.

the polar layered deposits, (c) clastic dikes formed from sand that was squeezed or injected upwards into the polar layered deposits from below, or (d) igneous dikes, composed of solidified magma (molten rock) injected into cracks in the polar layered deposits. In any case, the ridges were exposed to the air as wind scoured away the less-resistant polar layered materials."

Unconventional Mars investigators have always looked at the Inca City as potentially artificial, citing their resemblance to partially buried ruins on the Earth. Ancient structures like the ruins near Ashur, Iraq bear a striking resemblance to the Inca City.

Viking 2 images of the region were inconclusive, but in 1998 and again in 2002, the *Mars Global Surveyor* spacecraft got newer and better resolution images of the Inca City and the area around it than ever before.[2] What they revealed was that the impression of buried, rectilinear formations very much like partially buried ruins on Earth was not merely an optical illusion. A spectacular close-up of the ridges revealed that whatever was causing the uplifts was completely buried under a layer of sand and CO_2 frost. They also revealed a new mystery.

The image 75 feet per pixel image, MOC-7908, showed not just the underlying structure of the Inca City, but it also showed

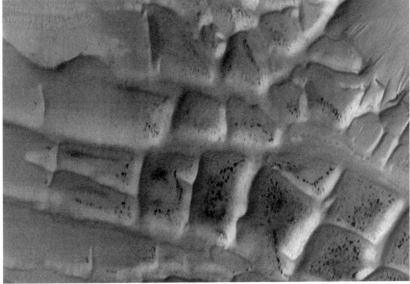

Mars Global Surveyor close-up of the Inca City "ridges" and the strange black spots.

87

some unexpected pitch-black "leopard spots" on the ground inside the "cells." No one, not even NASA, could venture an explanation for them. "The dark spots are enigmatic," the press release said. "The spots are about 20 to 100 meters (65 to 327 ft) in size. They might be large boulders or clusters of gravel and cobbles that were just warm enough [when the picture was taken] that they did not retain the ice/snow on their surfaces. Alternatively, they could be small, dark sand dunes." What they look like of course, is some sort of seepage, possibly even oil, from below the ground. These "leopard spots" have been seen elsewhere on Mars, and to this day there is no viable explanation for them.

Critics jumped on the new images, claiming that they "proved" the Inca City was not artificial. This was more than a bit of a stretch, since whatever caused the ridges was obviously still buried and they were remarkably similar to buried artificial structures on Earth, like the Khorezmian fortress at Koy-Krylgan-kala in Uzbekistan.

Khorezmian fortress at Koy-Krylgan-kala in Uzbekistan

When it was first overflown in the 1950s, it had been assumed to be an impact crater because of its circular form and buried ridges. Subsequent excavation by archeologists showed it to be an extensive circular structure that had been buried by the sands of time. To expect ancient, highly eroded archaeological "ruins" on another planet to look as clean as a recently excavated dig on Earth is not only naive in the extreme, it's not even logical. Even though the Inca City is on a much larger scale and completely unexcavated, this does not mean that artificiality could be ruled out *a priori*— certainly not based on a single MOC image that actually confirmed the high strangeness of the area.

Then in 2002, the *Mars Global Surveyor* got another, broader context look at the Inca City, and the mystery deepened...

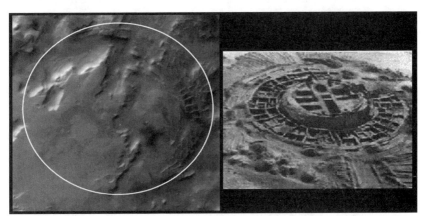

Mars Orbiter Camera image E09-00186 (L) and the Khorezmian fortress (R).

What the new image showed was that the Inca City was not an isolated formation. It was part of much larger, circular structure that bore an eerie resemblance to the Khorezmian fortress. The press release that accompanied the image acknowledged the obvious.

"The Mars Global Surveyor (MGS) Mars Orbiter Camera (MOC) has provided new information about the 'Inca City' ridges, though the camera's images still do not solve the mystery. The new information comes in the form of a MOC red wide angle context frame taken in mid-southern spring... The MOC image shows that the 'Inca City' ridges, located at 82°S, 67°W, are *part of a larger circular structure* [emphasis theirs] that is about 86 km (53 mi) across. It is possible that this pattern reflects an origin related to an ancient, eroded meteor impact crater that was filled-in, buried, then partially exhumed. In this case, the ridges might be the remains of filled-in fractures in the bedrock into which the crater formed, or filled-in cracks within the material that filled the crater. Or both explanations could be wrong. While the new MOC image shows that the 'Inca City' has a larger context as part of a circular form, it does not reveal the exact origin of these striking and unusual Martian landforms."

Again, the comparison to artificial, archeological ruins is strong. Just like the Khorezmian fortress, the initial idea is that it may be an impact crater. The problem is that the Inca City and the larger circular structure it is a part of doesn't look anything like an impact crater—but it looks exactly like artificial ruins here on Earth.

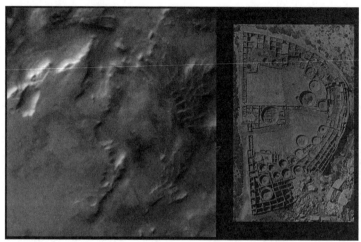

Comparison of the "Inca City" to the Anasazi structures in Chaco Canyon.

The mystery of the Inca City certainly can't be resolved from orbital images. The only way to know for sure at this point whether it is an artificial ruin or the most exotic natural formation in the solar system is to land there and excavate. But if NASA will ever go to that much trouble, there are several places they might want to land and excavate first...

THE MONUMENTS OF MARS

The possible existence of Ancient Alien artifacts on Mars actually got its initial push into the mass consciousness several years before Sagan's "Pyramids of Elysium" were popularized. It happened in 1976 when one of the *Viking* orbiters took a single, extraordinary picture of the surface of Mars.

Compared to *Mariner*, the *Viking* missions of the mid-1970s were wildly ambitious in their scope. Having given up on the idea of finding anything but the simplest microbial life on Mars, NASA designed two basically identical missions to operate side by side and search for such life after their arrival in June, 1976. The *Viking* missions actually consisted of four vehicles—two landers and two orbiters grouped together and called *Viking 1* and *Viking 2*, respectively. The orbiters contained upgraded cameras from the *Mariner* missions, and were to map the surface at high resolutions in order to find appropriate touchdown locations for their twin landers. The landers were then to separate from the orbiters and descend to the planet's surface to test for signs of life and take the first pictures from the Martian landscape. The first month in orbit was spent mapping the Martian surface at resolutions of 150 to 300 meters per pixel, although some selected areas (the possible landing sites) were imaged at spatial resolutions as high as 8 MPP. It was during this process that something extraordinary happened that threatened to change the canonical NASA perspective that Mars was a cold, dead world bereft of life for many billions of years.

On July 25, 1976, a project scientist at NASA's Jet Propulsion Laboratory named Toby Owen was looking at images of potential

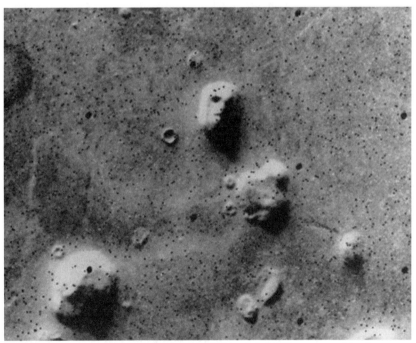

The "Face on Mars" (Viking frame 35A72) as it initially appeared in the press, July 25, 1976.

landing sites in a region known as Cydonia when he put a magnifying glass over *Viking 1* orbiter frame 35A72 and exclaimed "Hey, look at this!" What got him so excited was what appeared to be nothing less than a human face, sculpted into the Martian sands and staring back at him.

The next day, NASA held a daily press briefing in which "The Face on Mars" was the clear highlight. Dr. Gerald Soffen, a *Viking* project scientist, addressed the assembled press, including Richard C. Hoagland, my co-author on *Dark Mission*, who was covering the *Viking* missions as a science reporter for *American Way* magazine and CBS News. Soffen introduced the Face image with the statement "Isn't it peculiar what tricks of light and shadow can do...? When we took another picture a few hours later, it all went away; it was just a trick, just the way the light fell on it." Although the Face made newspaper headlines all over the world the next day, no journalist, including Hoagland, took it seriously. They all accepted NASA's explanation that there were disconfirming photos taken later that same Martian day. But curiously, neither Soffen, Owen nor any other NASA scientist showed anyone in the press room the second, disconfirming photograph which they flatly claimed existed.

That's because there wasn't one.

Because of various factors, there couldn't have been. For one thing, the first image was taken late in the Martian day, around 6 PM local Cydonia time. The shadows were already quite long and in any event, the *Viking* orbiter was on a 24.66 hour polar orbit (24 hours, 39 minutes, 36 seconds) compared to Mars' Sidereal rotation period of 24.622 hours (24 hours, 37 minutes, 19 seconds) a difference of 2 minutes and 16 seconds. Given that Mars' orbital rotation speed is about 540 miles per hour at the equator, that means that the next orbit—over a day later, would have taken the *Viking 1* orbiter between 15 and more than 20.4 miles away from the Face. And because of that there was no way it could have been anywhere near Cydonia "a few hours later," unless you consider more than a full day to be "a few hours." Not to mention that, "a few hours later," it would have been pitch-dark at Cydonia and the site of the Face anyway. In fact, given the speed of *Viking 1's* orbit, it's doubtful that another orbit had even been completed by the time the press briefing was held. Finally, the image number "35A72" indicates that it was taken by orbiter "A" (*Viking 1*) on orbit 35, and was the 72nd picture taken on that day. The only other good image of the Face from the *Viking* era is 70A13, the 13th picture taken by *Viking 1* fully 35 orbits (days) later. The bottom line is that there is no way the "disconfirming image" alluded to by Soffen could have ever existed.

So why did they rush to get it out in the press and claim that there was nothing to this "Face" thing? And how could the *Viking* project scientists have made such a huge mistake in their claim about a second, disconfirming image?

They didn't. They lied.

And they continued to lie about it for almost two decades thereafter.[3] In fact, it wasn't until an independent science evaluation by Dr. Stanley McDaniel in his study named *The McDaniel Report* was published in 1993 that NASA stopped sending out misleading letters to congressmen who inquired about Cydonia. Up until that time, all of NASA's responses to official congressional inquiries included the "second disconfirming photo" claim, which they had to know was false. It's important to note NASA never actually *admitted* no such photos existed (which they could not have), they just eventually stopped claiming that they *did*.

None of this should be too surprising considering the evidence we uncovered in *Ancient Aliens on the Moon* and *Dark Mission* of image tampering by NASA. There's a reason the beat reporters covering the agency in the 1960s used to joke that NASA stood for "Never A Straight Answer," and why President Reagan's science advisor, George Keyworth, once said of the agency in testimony before Congress: "All government agencies lie part of the time, but NASA is the only one I've ever encountered that does so routinely."[4]

So why then all the fuss about the Face if it was really a simple "trick of light and shadow?" The reason may have had something to do with why frame 35A72 was taken in the first place; they were thinking about landing there.

The *Viking 1* lander had put down on July 20, 1976 in the

First picture from the surface of Mars, July 20th, 1976, from *Viking* 1.

Chryse Planitia region of Mars, just a few days before the "Face" picture was released on July 26th. Sending back photograph after photograph of the planet's surface, the *Viking 1* lander immediately began taking revealing panoramas of the landing site. Cydonia had been selected as the landing site of the second of the *Viking* landers, *Viking 2*, but within a few days of that first "Face" image (35A72), rumblings began about changing the *Viking 2* landing site.

The Face image must have caused quite a bit of consternation at NASA's Jet Propulsion Laboratory, because in many ways, Cydonia (designated landing site B-1, 44.3°N, 10°W) was a better choice than Chryse Planitia as a landing site. Cydonia had been chosen as the *Viking 2* prime site because it was low, about five to six kilometers below the mean Martian surface, and because it was near the southernmost extremity of the wintertime north polar hood. B-1 also had the advantage of being in line with the first landing site, so the *Viking 1* orbiter could relay data from the second lander while the second orbiter mapped the poles and other parts of Mars during the proposed extended mission.

Cydonia was also considered a good spot to find water, which would have made it an ideal place to search for microbial life, but *Viking* project scientist Hal Masursky expressed concerns about the geology of the region. He asked David Scott, who had prepared the geology maps, to work up a special hazard map for the Cydonia B-1 site. It was almost as if they were looking for an excuse to dash the Cydonia landing. After studying the map, Masursky came to the conclusion that the area was not "landable." This analysis was evidently made with maps based on *Mariner 9* photographs. He told NASA project scientists Tom Young and Jim Martin, however, that there was one hope: wind-borne material may have mantled the rough terrain and "covered up all those nasties we see."[5]

So the ostensible reason for changing the targeted landing site was that Cydonia was suddenly considered "too rocky" for the *Viking* lander to risk a touchdown. It was further claimed that the "northern latitude" of Cydonia was partly to blame for this rough surface, and a more suitable landing site should be sought farther south. In the end, the *Viking 2* lander set down in a region known as Utopia Planitia, an even more northerly and rocky site than B-1 in Cydonia.

Nobody thought much of the venue change at the time, but since their new choice for a landing site contradicted their reasons for dropping Cydonia, it seemed that somebody at JPL was nervous enough about the Face to make sure that the *Viking 2* lander stayed well away from it. One NASA consultant, nonplussed by the odd flip-flop on the landing site, compared the choice to landing in the Sahara desert on Earth to look for life, rather than in a more hospitable climate. "A multimillion dollar effort may have overlooked 'pay dirt' and may have become a trivial event. . . . A poor selective factor had been used to choose an area of minor geological and biological significance. It was like choosing the Sahara Desert as a suitable landing site on our own planet."[6] In an even more bizarre decision, NASA took two more high resolution images of Cydonia—70A11 and 70A13—in mid-August, well after they decided the region was unsuitable for a landing. In doing so, they sacrificed precious orbiter resources that could have been used to photograph another presumably more suitable region of Mars. Had they seen something in 35A72 that made them curious?

Things remained pretty quiet on the Cydonia front after that until 1979, when a pair of imaging specialists at NASA's Goddard Space Flight Center, Vince Dipietro and Greg Molenaar, decided to look for the Face. They quickly found 35A72 (labeled simply "Head" in the *Viking* image files) and their early enhancements seemed to argue against the "trick of light and shadow" explanation. The more they worked with it, the more it looked like a sculpted face. They then began to look for other possible images of the Face from other orbits. They were surprised to find both that potentially interesting images of the Face taken on subsequent orbits seemed to have disappeared, and there seemed to be no trace of the "disconfirming photographs" that Gerald Soffen had alluded to five years earlier. After an exhaustive search of the *Viking* archives, they discovered a second *misfiled* Face image, 70A13, taken 35 orbits later at a 17° higher sun angle. They never did find the supposed "disconfirming" image and they were the ones who subsequently established that since the next *Viking* orbit took it well away from Cydonia and was at Martian nighttime, no such image could conceivably exist.

They then began to seek input from other scientists. Although

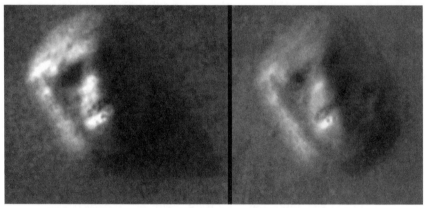

Processed Viking images 35A72 (L) and 70A13 (R)—(Carlotto).

stymied in their attempts to get articles on the Face published in the peer-reviewed journals, Dipietro and Molenaar eventually managed to get some of their enhancements of the Face into the hands of Richard C. Hoagland. Although Hoagland had requested the prints in order to study the image enhancement technique being used by Dipietro and Molenaar rather than the Face itself, he was intrigued by what he saw. The enhancement tool, programmed by Dipietro and Molenaar themselves, was called "S.P.I.T.," Starburst Pixel Interleave Technique—an early version of image interpolation tools now commonly used in image enhancement. What these early tools showed (long before there was anything like Photoshop available) was that the more the images of the Face got clearer and cleaned up, the more it *looked* like a face. After some discussions with the two scientists, Hoagland was able to secure funding for the first Independent Mars Investigation ("IMI") under the auspices of the Stanford Research Institute.

From the beginning, Hoagland realized that the question of the Face required special consideration. As far as any members of the IMI knew, no one had ever attempted such an investigation before, and there were therefore no set rules as to how the "Face problem" should be approached. Working from the idea that if the Face were indeed artificial it would likely be beyond the experience of geologists and planetary scientists, Hoagland determined that the research required a group with a broad cross section of skills from the various "hard" and "soft" sciences. This approach allowed the original members of the IMI to examine the Face from every possible scientific perspective, and to cross-reference their results with a ready-made peer review panel. What they found only deepened the mystery. After close scrutiny of both 35A72 and 70A13, some initial conclusions were immediately evident.

Since the Face was not a profile view as seen in terrestrial

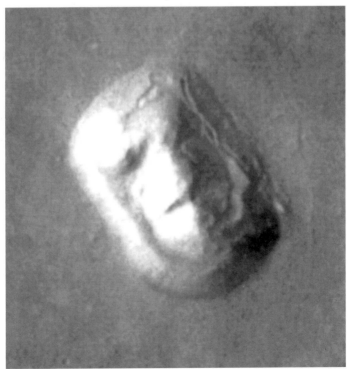

The Face on Mars from Mars Odyssey 2001.

rock faces like the Old Man in the Mountain in New Jersey, but rather a direct, overhead view more akin to the presidential monument at Mount Rushmore, the members of the investigation quickly discounted the idea they were just "seeing things." The Face seemed to have specific primary and secondary characteristics of human faces, including brow ridges, eye sockets, a full mouth and even a nose. The higher sun angle image 70A13 showed that the beveled platform upon which the Face was seated appeared to be in the range of 90% symmetrical, despite the presence of a data error in the image that distorted the area around the eastern "jaw."

This second image also confirmed the presence of a second eye socket on the shadowed eastern side, and that the level platform upon which the facial features rested was uniform in height and symmetrical in layout at least as far down as the mouth. This image also eventually revealed (under later enhancements by Dr. Mark Carlotto) what appeared to be teeth in the mouth, bilaterally crossed lines on the forehead and lateral striping on the western half. Both images also showed a mark of some kind, dubbed the "teardrop," on the western side of the face just below the eye socket.

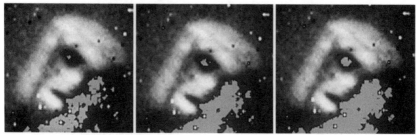

"Bit slice" enhancement of the western eye socket of the Face on Mars by Vince Dipietro.

Later, using a new "bit slice" imaging technique, Dipietro found what he claimed was a spherical "pupil" in the western eye socket. It will be important later to remember that the critics of the investigation, among them Dr. Michael Malin of Malin Space Science Systems (who controlled the camera for the *Mars Global Surveyor* probe), claimed that the "pupil" was not really there and was beyond the resolution limits of the data.

In fact, to this day, that is what most NASA and NASA funded scientists continue to claim. They argue that the Face is an example of something they call "Pareidolia," a mythical, made-up tendency for humans to see Faces and facial patterns where none exist. For the record, there is no such thing as "pareidolia." It is a phony, pseudo-scientific term invented in 1994 by a UFO debunker named Steven Goldstein in the June 22nd, 1994 edition of *Skeptical Inquirer* magazine,[7] which should tell you all you need to know about its credibility in the realm of ideas. Despite a complete lack of any valid scientific studies on the supposed "phenomenon," it is still commonly cited by debunkers like James Oberg and Phil "Dr.

"Pareidolia," or Prosopagnosia?

99

Phil" Plait to give an academic air to their knee-jerk dismissal of the Cydonia anomalies. Of course, to demonstrate their point, they always show the original, crude and unprocessed *Viking* image of the Face on Mars rather than any of the multitude of clearer, better images taken in the decades since.

There is however another very real human tendency that unlike the mythical "pareidolia," is actually an extremely well-documented and medically established disorder—Prosopagnosia. Simply put, Prosopagnosia is a brain disorder that renders the poor souls that have it completely unable to recognize faces when they see them. According to some medical studies, as much as 2.5% of human population may suffer from this disorder, and apparently a disproportionate number of those afflicted have found jobs in the NASA planetary science community.

It should also be noted that advocates of pareidolia have recently changed their definition of it from specifically mentioning facial recognition to just a supposed human tendency to see "patterns" where none exist. This is much like the man-made global warming believers who no longer refer to "global warming" but

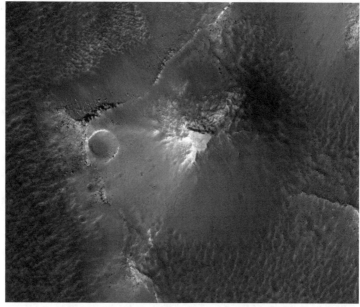

A Meso-American style pyramid in the "City" area at Cydonia. The scale is on a par with the Great Pyramid at Giza.

rather to a generic "climate change," because the data has piled up proving that no such thing as man-made global warming actually exists.

By the mid-1980s, Dr. Mark Carlotto, an imaging specialist, had been brought in to the second Mars research group organized by Hoagland, the Mars Investigation Group. He used new imaging techniques to bring out more detail than Dipietro and Molenaar's earlier method had from the two *Viking* images. As the enhancements got better and better, other objects around the Face began to jump out of the data. Like the Face, they were inexplicable and mysterious.

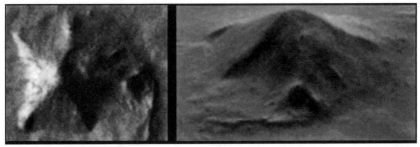

The pentagonal D&M pyramid of Mars in overhead (L) and 3D perspective (R) views (Carlotto).

Hoagland was the first to realize that all of this detail was ultimately meaningless if it turned out that the Face was an isolated landform. No matter how much it looked like a face, if it was all by itself, with no evidence of any civilization around it to have constructed the monument, then it could simply be a marvelous trick of erosion and shadow after all.

So Hoagland and the members of the investigation began to look in the immediate vicinity of the Face to see if there was any other evidence of anomalous objects nearby. Dipietro and Molenaar had previously noted a cluster of "pyramidal" mountains to the west of the Face, and they had also pointed out a massive object (1.5 km in height) to the south that appeared to be a four-sided pyramidal mountain. Hoagland dubbed this cluster of mountains the "City," and the massive pyramidal mountain the "D&M Pyramid," in honor of Dipietro and Molenaar. Enhancements by Carlotto revealed that the "D&M" seemed to be a five-sided *pentagonal* object, rather than four-sided, as Dipietro had argued.[8] Carlotto's enhancements also showed that there was a bulge on the northern-right facet of the D&M and an "entrance wound" of sorts where something had appeared to penetrate the object from outside. The "City" objects displayed a number of unusual geomorphic characteristics as well. In time, features like the "City Square" (an arrangement of equally spaced mounds with a direct sightline to the Face), the "Fortress" (an object just outside the "City" which seemed to have a triangular shape and two straight walls), the "Tholus" (a rounded mound which closely matched man-made earthen mounds in England in shape and layout—complete with a "trench" around it), the "Cliff" (a long, almost perfectly straight ridge atop what appeared to be a platform built over the ejecta from a nearby impact crater) and the "Crater Pyramid" (a tetrahedral pyramidal mound somehow perched on the rim of the impact crater) formed what became known as the "Cydonia Complex."

Further examination provided additional details. There was evidence of digging next to the Cliff, implying that the platform

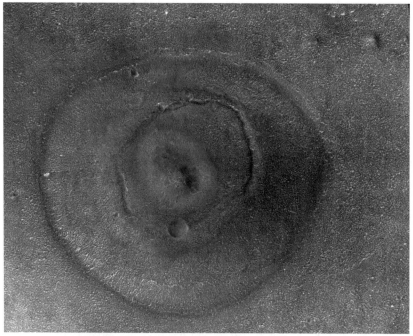

The "Tholus" lies just a few miles from the Face on Mars.

upon which it rested had been built up from this material. The Tholus turned out to have an "entrance" of sorts at the top, a walkway that went from the base to this entrance and an eroded, pyramidal cap on it. The D&M had what appeared to be almost a bottomless crater next to it, and the right side of the object seemed to bulge out slightly, as if from an internal blast (caused by whatever made the crater?). The City turned out to have a degree of organization to it, and architect Robert Fiertek did an extensive reconstruction of the original layout.

By the mid-1980s, the various members of the investigation were ready to present their findings to the scientific community and call for more analysis and better pictures to determine the validity of their observations. They met with a chilly reception.

Efforts to get their work published in peer-reviewed journals were quickly rebuked. Members later found out that in most cases, the papers were rejected without even having been read, much less "reviewed by peers." Behind-the-scenes efforts to get assistance from prominent members of the scientific community met with a bit more success, as Carl Sagan helped Carlotto get a couple of papers published in computer optics journals and featured his images in an update of the *Cosmos* TV series on DVD. Oddly, at the same time he was doing this, Sagan was attacking the whole issue publicly with an infamous disinformation piece in *Parade* magazine. This would not be the last time that Sagan contradicted himself on Cydonia.

Attempts to present their data to peers directly, via scientific conferences and the like, were also met with resistance. When members of the Mars Investigation Group presented a poster session and a paper at the 1984 "Case for Mars" conference, they were surprised to find later that their presentation and their paper had been expunged from the officially published record of the conference, as if they had never been there.

Undaunted, Hoagland and the others continued their research. Yet, as documented by Dr. Stanley V. McDaniel of Sonoma State University in his voluminous *The McDaniel Report*, NASA seemed to have an aversion to investigating what seemed to be an ideal subject for the agency's agenda. In fact, they vociferously refused to even consider making the imaging of Cydonia a priority for any new Mars missions, like the upcoming *Mars Observer*. Beyond that, they continued to insist, in response to inquiries from congressional leaders and the public, that the non-existent "disconfirming photos" proved that the Face was just an optical illusion. Only after many years (17) of repeatedly pointing out to NASA that no such images existed and under heavy pressure from Senator Diane Feinstein, did they finally cease making this claim.

Dr. Carlotto moved the research in a new direction when he developed a fractal analysis technique (to discern which objects in

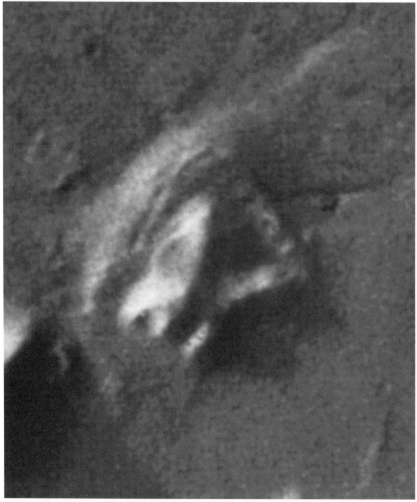

The "Fortress" from *Viking* frame 35A72.

an image were the least consistent with the "natural" background) to be used on the Cydonia images.[9] After an initial study of about 3,000 square kilometers around the Face, Carlotto and his partner, Michael C. Stein, determined that the Face and the Fortress were the two most "non-fractal" objects in that terrain. Pressed to go even further, they eventually used the program on images covering some 15,000 square kilometers around the Face. The results were consistent with the earlier run-through. The Face was by far the most non-natural object in the surveyed terrain. NASA responded through Dr. Malin to the effect that Carlotto had not measured anything other than the fact that the Face was different, not necessarily artificial, and suggested that if he applied the technique to a broader area, he would find that the curve would smooth out, and that the Face was not all that unusual.

This response ignored the fact that Carlotto had already done just that by expanding the survey from 3,000 square kilometers to 15,000 and that, contrary to Malin's assertions, the Face's uniqueness was even more pronounced. Lacking the funds to expand the research even further, Carlotto offered to turn the program over to NASA so that the agency could continue the survey over the entire Martian surface. NASA's response was a polite "thanks, but no thanks."

Up to this point, a lot of the behavior of NASA and the planetary science community could be viewed through the tint of simple prejudice or ignorance. No one wanted to be the next Percival Lowell, sticking their chin out on the issue of life on Mars only to have their reputation forever soiled if the data turned out to be wrong. Other members of the broader scientific community simply refused to even consider the possibility that Mars was once inhabited by an advanced civilization, no matter what the evidence. Their models and training had taught them that Mars was a cold, dead world, and had been for billions of years. The notion that someone had been there, built these monuments and then left sometime in the distant past was just too destabilizing to their way of thinking.

The next step in the investigation was even more radical however, and it is here that NASA's resistance turned to active disinformation and suppression.

(Endnotes)

1 A Type II civilization refers to the Kardashev scale proposed in 1964 by Soviet astronomer Nikolai Kardashev. According to this scale, humanity is defined as a Type I civilization. A Type II civilization is defined as "A civilization capable of harnessing the energy radiated by its own star (for example, the stage of successful construction of a Dyson sphere), with energy consumption at $\approx 4 \times 1033$ erg/sec. Lemarchand stated this as "A civilization capable of utilizing and channeling the entire radiation output of its star. The energy utilization would then be comparable to the luminosity of our Sun, about 4×1026 watts." Kardashev, Nikolai (1964). "Transmission of Information by Extraterrestrial Civilizations". Soviet Astronomy (PDF) 8: 217. Bibcode:1964SvA.....8..217K

2 http://www.msss.com/mars_images/moc/8_2002_releases/incacity/

3 *The McDaniel Report--On the Failure of Executive, Congressional, and Scientific Responsibility in Investigating Possible Evidence of Artificial Structures on the Surface of Mars and in Setting Mission Priorities for NASA's Mars Exploration Program,* North Atlantic Books (June 1993), ISBN-13: 978-1556430886, pp. 13.

4 Congressional record, March 14, 1985

5 http://history.nasa.gov/SP-4212/ch10.html

6 http://www.dudeman.net/siriusly/cyd/vik.html

7 http://www.wordspy.com/words/pareidolia.asp

8 DiPietro, V. and Molenaar, G., *Unusual Martian Surface Features*, Mars Research, Glenn Dale, MD, 1982

9 http://carlotto.us/martianenigmas/Papers/JBIS1990.pdf

CHAPTER 4
THE MESSAGE OF CYDONIA

Early on in the Cydonia investigations, Richard Hoagland had proposed that there might be a broader, contextual relationship between the various landforms identified as anomalous. By themselves, the Face, Fort, City, Tholus, the Cliff, the Crater Pyramid and the D&M Pyramid were highly unusual landforms that were incongruous with the existing geologic model of Cydonia. But Hoagland had also noted several interesting mathematical and geometric relationships between the potential monuments on the Cydonia plain. He noticed, for instance, that the three northward edges of the pentagonal D&M seemed to point to other key features of the complex. Using orthographically rectified images provided by the US Geological Survey and the Rand Corporation, he drew lines defined by these faceted edges across the images of Cydonia. One edge passed right through the center of the City Square, the next right between the eyes of the Face, and the next straight across the apex of the Tholus. He also noted several "mounds" in and around the City. They were consistent in terms of size (about the scale of the Great Pyramid at Giza) and shape, and also seemed to form a perfect equilateral triangle.

It is important to understand the sequence in which these observations were made. Hoagland has often been accused of "circular reasoning," of just drawing lines on the photos until they "hit" something and then declaring that object to be a "monument." This is not, in fact, the case.

As we have seen, and has been well documented by Hoagland, Carlotto, Pozos, McDaniel and others, the observations of the anomalous geomorphic characteristics came first. It was only later, when some thought was given to how potentially artificial objects might have a contextual relationship, that the measurements were

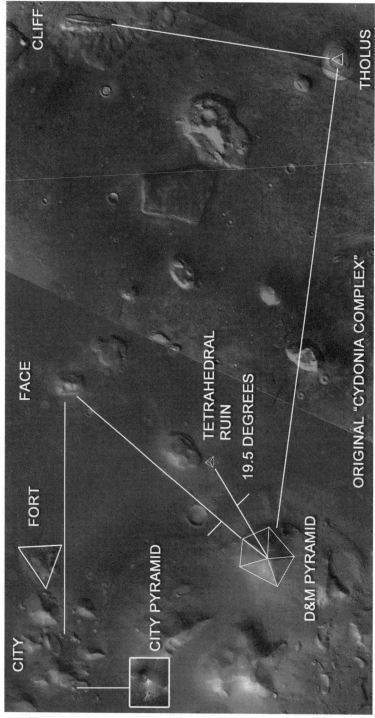

The original "Cydonia Complex," with key features identified.

made. Even then, the methodology could have become "circular" if certain precautions were not taken. Hoagland carefully used only techniques that had been previously established by archeologists in their surveys of ancient ruins.[1]

Taking a page out of the SETI manual, Hoagland decided that any intended message would almost certainly have been inscribed more than once. If an architect were seeking to send a clear mathematical signal to a civilization that might happen upon his creation, he would surely have reinforced the message, since a single mathematical relationship could not be distinguished from random noise. So a cornerstone of the whole process was that any "significant" mathematical relationship must occur redundantly. He also made certain not to include any object that was not significant in some other way to the model. If an object was not anomalous in any way, but stood at a significant location in the alignment model, it was rejected. Each and every relationship that would be considered significant had to be a candidate for inclusion on at least two grounds.

A prime example of this is the City Square. It was originally considered a potential candidate for artificiality because of the way the four mounds were equally spaced around a central nexus. Additionally, the four mounds seemed to be almost identical in height, scale and volume. So the fact that the center of the City Square was later found by Hoagland to lie along a direct line marked by the northwest facet of the D&M was only significant because of these previous observations. Without the initial geomorphic issues calling the natural origin of the features into question, the later determined alignment would have been meaningless in Hoagland's methodology.

Yet he still faced a significant degree of criticism from "reductionists" inside NASA. The reductionist method seeks to isolate each and every data point in a given argument and break it down without reference to the greater context. Hoagland argued that this isolationist approach could not be valid in an investigation such as this one, since there would likely have been some form of intent in the mind of any "Martian architect," just as there was in all

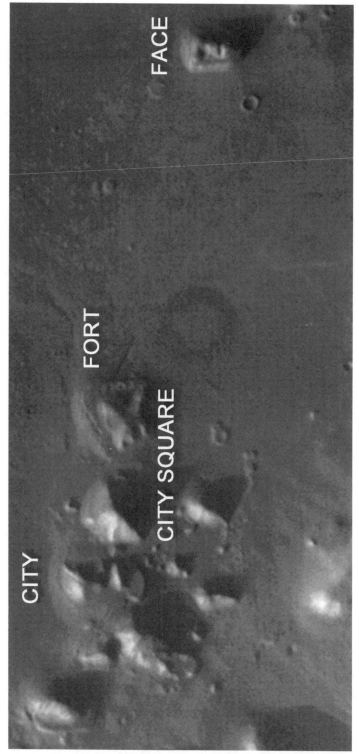

Viking image 35A72 showing the area of the City and the so-called "City Square."

earthly monumental architecture.

Interestingly, this was not the first time that someone had faced this sort of criticism from NASA.

On November 22, 1966, NASA released a *Lunar Orbiter* 2 image from the Moon in the vicinity of the crater Cayley B in the Sea of Tranquility (see the "Blair Cuspids" discussion in *Ancient Aliens on the Moon*). In it, there were objects casting extremely long shadows that seemed to imply that the objects themselves were towers of seventy feet or more. William Blair, a Boeing anthropologist, noted that the "spires" had a series of contextual, geometric relationships with each other. "If such a complex of structures were photographed on Earth, the archeologist's first order of business would be to inspect and excavate test trenches and thus validate whether the prospective site has archeological significance," he was as saying quoted in the *Los Angeles Times*.

The response from physicist Richard V. Shorthill (then of the Boeing Scientific Research Laboratory but later NASA) was swift and eerily reminiscent of the criticism aimed at Hoagland. "There are many of these rocks on the Moon's surface. Pick some at random and you eventually will find a group that seems to conform to some kind of pattern." This was an example of reductionism in its purest

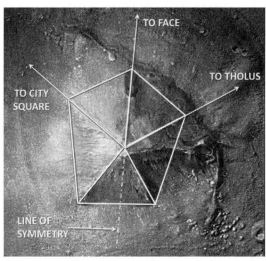

The D&M Pyramid of Mars, as reconstructed by Erol Torun of the Defense Mapping Agency in 1989 (ESA)

111

form.

Blair's rebuttal would later put the reductionist arguments in their appropriate context: "If this same axiom were applied to the origin of such surface features on Earth, more than half of the present known Aztec and Mayan architecture would still be under tree- and bush-studded depressions—the result of natural geophysical processes. The science of archeology would have never been developed, and most of the present knowledge of man's physical evolution would still be a mystery."

Fortunately, there were some interested researchers that were more open minded than the likes of Shorthill and NASA.

In 1988, Hoagland was approached by a man named Erol Torun—a cartographer and satellite imagery interpreter for the Defense Mapping Agency—and probably the most uniquely qualified person on the planet to render a judgment on the potential artificiality of the Cydonia enigmas. After attaining a degree in geology with a specialty in geomorphology, he had spent more than ten years of his professional life looking at remote imagery just like the original *Viking* data and distinguishing artificial structures from naturally occurring landforms. His analyses were used by the Defense Mapping Agency (later NIMA) to find camouflaged targets for military assaults in foreign countries.

After reading Hoagland's 1987 book *The Monuments of Mars*, he had written Hoagland expressing his surprise that his initial assumptions about the subject were not supported by his subsequent analysis. He was particularly impressed with the geometry and geology of the D&M Pyramid. "I have a good background in geomorphology and know of no mechanism to explain its formation," he wrote Hoagland. He was fascinated by its near perfect five-sided pentagonal form, its bilateral symmetry about a central axis (which pointed directly at the Face) and the fact that it stood alone and bore no resemblance to other nearby landforms, especially the ones considered potentially artificial. His initial analysis of the D&M ruled out a variety of natural mechanisms including water (fluvial) processes, mass wasting, crystal growth and volcanism. But the main explanation NASA has used over the years—wind—was also quickly dismissed as a possibility.

Viking 1 landing site from current NASA imagery compared to original data. Note the
difference between official NASA "Technicolor red" and actual Arizona-like landscape.

Comparison of Mars rock from Viking 1 landing site with greenish-blue patch (L) and lichen-covered rock on Earth (R).

The D & M Pyramid back side.

The D & M Pyramid front side.

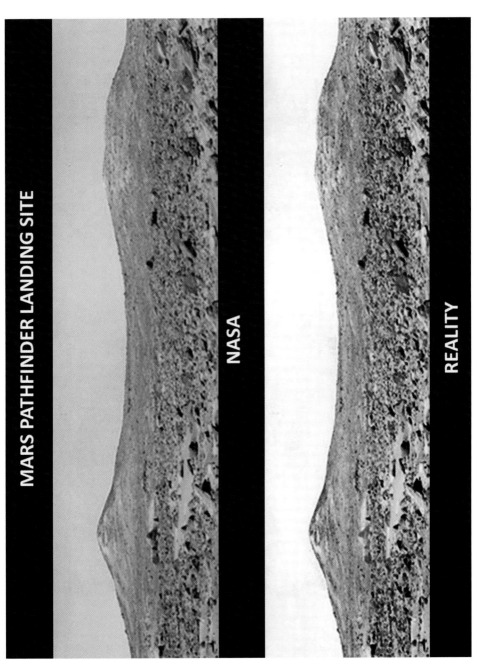

Official NASA panorama of Mars Pathfinder landing site (top) and color correct-
ed version (bottom) with blue sky.

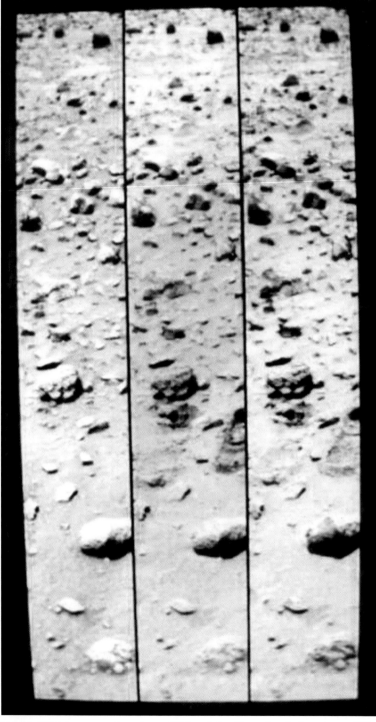

AS14-66-9301 Images taken from Viking 1 landing site during three different seasons in Martian year showing growth of lichen-like blue-green patches on rocks in the area. Image courtesy Dr. Gil Levin.

MOLA topography map of Mars showing cratered highlands (red) and cue ball smooth lowlands (blue) and the distinct "Line of Dichotomy" encircling the planet/moon. (NASA)

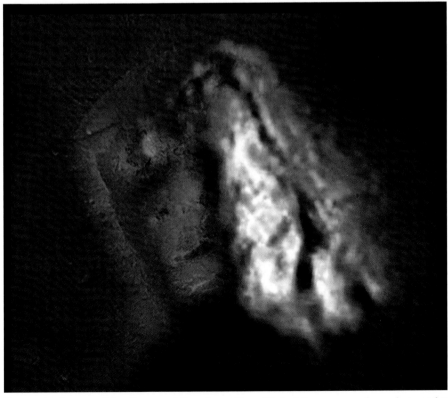

Composite image of the Face on Mars from Mars Global Surveyor and pre-dawn color image from Mars Odyssey showing underlying, reflective, mechanical/structural aspects of the Face. Image courtesy Richard C. Hoagland.

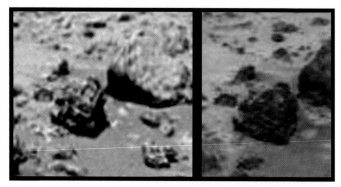

Side-by-side comparison of the same "rocks" from European Space Agency (L) data and NASA (R). Note that what appears to be mechanical debris in the ESA data is patched over in NASA version, and other mechanisms are optically removed by NASA.

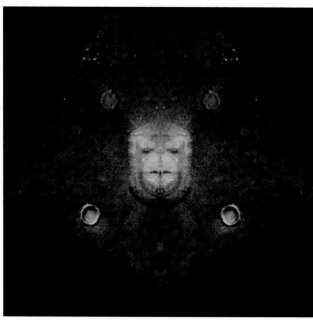

The Face on Mars in a montage.

Sphinxes guarding pyramids on two different planets.

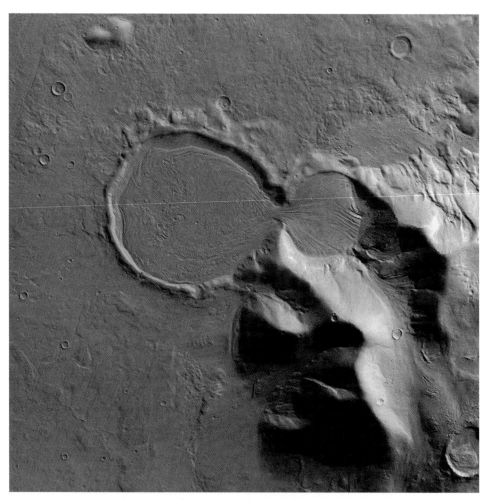

The hourglass-shaped crater in the Hellas Basin, Mars. (NASA)

The Martian Sphinx with pyramid in the background.

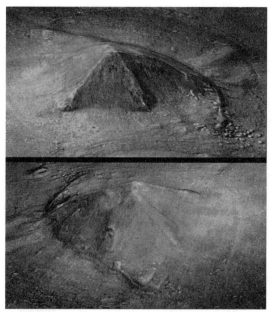

3D perspective views of the five-sided D&M pyramid, generated by *Mars Express* (ESA)

To put it simply, NASA is "all-in" on wind as an explanation for the shape of the D&M and other objects like the Elysium pyramids we looked at earlier. NASA's Jim Garvin, the chief scientist for NASA's Mars exploration programs from 2000 until 2004, even went so far as to claim on a National Geographic Channel TV special that the D&M and other objects like it were formed when the prevailing Martian winds magically rotated around the apex of the D&M every thousand years or so, forming a five-sided pyramid.[2] This baldly nonsensical claim had already been ruled out by Torun's analysis decades earlier. While Torun acknowledged in his report[3] that "Aeolian (wind) erosion is now the dominant mechanism of geomorphic change on Mars and has been since the disappearance of liquid water and the cessation of volcanic activity," he also pointed out that any such changing winds would erase the sharp edges previously formed before the change in direction. Torun also compared the D&M to wind eroded features that are commonly seen on both Mars and Earth— they are called "yardangs."

Even though according to Torun "Yardangs can occasionally

Terrestrial examples of wind formed yardangs. These 3 separate hills were once one much larger formation, but swirling winds cut through soft spots on the hill and formed this group of rounded, layered rock.

exhibit sharp edges, roughly flat sides, and bilateral symmetry," the simple fact is that with the winds changing randomly in the Martian environment, yardangs always form in closely spaced groups and end up as rounded, not multi-faceted, formations. Said Torun: "Locally reversed airflow can cut a flat surface perpendicular to the wind direction on the leeward side of a wind-cut hill. This locally reversed airflow, and associated surface level turbulence, would prevent the formation of this hypothetical five-sided [object]. Each time the wind shifted to a new direction, the reversed airflow would start erasing the edges formed by other wind directions. The end result would not be a pyramidal hill, but rather a round one."

As we discussed in reference to the Elysium pyramids, another factor working against the D&M being a natural landform is its isolation. "Comparison of the D&M Pyramid with landforms known to be yardangs immediately reveals some serious inconsistencies," Torun wrote. "The D&M Pyramid is an isolated landform with no other nearby objects exhibiting a similar shape and orientation. It is rare for yardangs to be found in isolation." Rare, as in no one has yet produced a single example anywhere on Earth that looks even remotely like the D&M.

Torun also looked at other kinds of wind-eroded natural structures on Earth. "Another type of Aeolian (wind-formed) landform that can be somewhat pyramidal in shape are known as ventifacts. Terrestrial ventifacts are normally formed from small rocks that are exposed to the abrasive action of sand carried by the wind. Large ventifacts can also exist, produced from boulders and assuming a roughly pyramidal shape with *three* edges..." But when applied to the D&M, this explanation didn't fit either.

"Five-sided symmetrical ventifacts or yardangs appear to be totally nonexistent on Earth and Mars," Torun acknowledged. "Prevailing winds are not likely to have shifted periodically with perfect symmetry and timing. Even if this seemingly impossible condition were satisfied, another factor (the reversed airflow mentioned above) would prevent such an object from forming."

He went on to make several decisive conclusions:

"The overall morphology of the D&M Pyramid, with its straight edges and flat surfaces in radial arrangement, is inconsistent with the morphology of aeolian landforms. The nearby Face shows no evidence of wind faceting, and there are no intervening objects between the Face and the D&M Pyramid to deflect wind. Also inconsistent is the presence of a flat-faced protuberance at the front of the object, a flat surface that should not exist at the leading edge of wind cut features such as yardangs or ventifacts. It is reasonable to conclude that aeolian processes cannot have produced the D&M Pyramid due to the lack of a plausible mechanism of formation, and the absence of similar landforms on Mars or Earth."

And finally: "The Geomorphic (natural) Hypothesis is thus left with no mechanism that can explain the formation of the D&M

Pyramid. This object's five-sided shape and bilateral symmetry is unlike any landform seen to date in this solar system, and even small-scale phenomena such as crystal growth cannot explain its morphology."

So much for Jim Garvin and NASA's silly "wind" explanation.

And Torun was not the only one to find such inexplicable objects on Mars. In more recent times, as more and more missions have imaged the Martian surface at higher resolutions, more geometrically eroded pyramidal formations have been discovered. One of the best of these is a six-sided (hexagonal) pyramid very similar to the D&M found by Gary Ligiere, who runs the "Mad Martian" website. He found the object, which is every bit as artificial as the D&M, on a *Mars Odyssey 2001* visual camera image taken a short distance from the Face area in Cydonia.

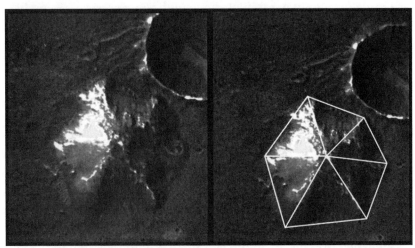

The hexagonal "Mad Martian" pyramid in Cydonia from NASA image V28082012.

But Torun had still more to add to his study of the D&M Pyramid. When he continued his analysis, what he found turned the Cydonia investigation upside down and sent it in a wholly new direction.

Torun had come to the Mars investigation as a skeptic, relatively certain he would find that the geomorphic interpretations and the early contextual alignments cited by Hoagland would turn out to be "false positives" in the search for answers to the riddle of

Cydonia. Yet once he had a chance to study the Cydonia images in detail, Torun concluded that the D&M itself was nothing less than the "Rosetta Stone" of Cydonia, finding a series of significant mathematical constants expressed in the internal geometry of the D&M. Being careful to avoid projecting his own biases on the measurements, Torun decided beforehand that he would restrict his analysis to just a few possible relationships.

As it turned out, not only did the D&M *have* a consistent internal geometry, it was also one full of rich geometric clues that spoke to him of a specific mathematical message. He found numerous repetitive references to specific mathematical constants, like e/π, √2, √3, √5 and references to ideal hexagonal and pentagonal forms. He also found geometry linking the shape of the D&M to other ideal geometric figures, like the Golden Ratio (Φ), the *Vesica Piscis* (which is the root symbol of the Christian church), and the five basic "Platonic Solids"—the tetrahedron, cube, octahedron, dodecahedron and icosahedron. Further studies found that the reconstructed pentagonal shape of the D&M, as determined by Torun

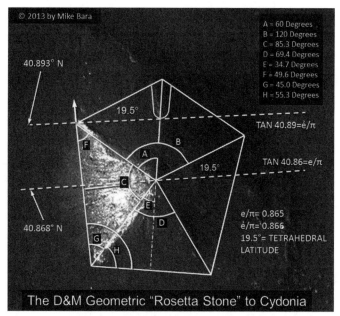

Internal Geometry of the D&M Pyramid. Adapted from Hoagland (2002) and Torun (1989).

117

before he took any of these measurements, is the *only one* that could produce this specific set of constants and ratios. More than that, these same constants showed up redundantly in all the different methods of measurement, and were not dependent on terrestrial methods of measurement (i.e., a radial measurement system based on a 360° circle). As Torun put it: "All of this geometry is 'dimensionless;' i.e. it is not dependent on such cultural conventions as counting by tens, or measuring angles in the 360 system. This geometry will 'work' in any number system."

Torun also discovered that the latitude of the reconstructed apex of the D&M was 40.868, which was a very close fit for the arctangent of the e/π ratio. Torun concluded that this was most likely an intended clue to anyone studying the structure that it was artificial. Such a self-referential numeric connection was, in his opinion, quite unlikely to occur by chance. Put simply, the D&M "knows where it is" on the surface of Mars.

After receiving Torun's study, Hoagland quickly realized that they were on the verge of a potentially important discovery. If Torun's numbers were repeatable throughout the Cydonia Complex, and if the same angles and ratios appeared in the larger relationships between the already established potential "monuments," then they would have a very strong argument that Torun's model was valid. Again being careful to only take measurements between obvious features like the apex of the Tholus and D&M, the straight line

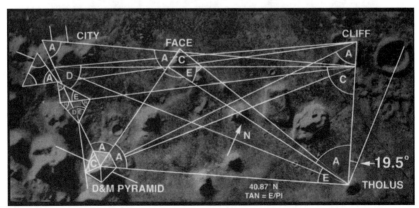

Hoagland's Cydonia "Geometric Relationship Model" – (Carlotto Enhanced Mosaic).

118

defined by the Cliff, the center of the City Square, the apex of the tetrahedral Crater Pyramid, Hoagland found that many of the same angles, ratios and trig functions applied all over the Cydonia Complex.

Somewhat stunned by what they had found, Hoagland and Torun had come to the realization that there was a message on the ground at Cydonia. The problem was that they didn't know what that message was trying to say.

In the message itself was the key to decoding it. One of the angles noted by Torun within the D&M was 19.5°, which occurred twice. Hoagland also found the same 19.5° encoded in the broader Cydonia complex three more times. Searching for the significance of this number, they eventually determined that it related to the geometry of the tetrahedron. The simplest of the five so-called "Platonic Solids" (because it is the most fundamental three-dimensional form that can exist), it made a certain kind of sense to

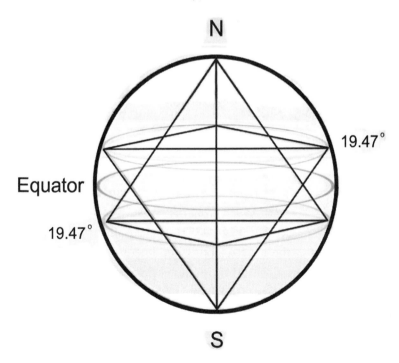

Diagram of a double tetrahedron circumscribed by a sphere. Corners will "touch" the inside surface of the sphere at 19.5 degrees.

119

use this "lowest order" geometric shape as a basis for establishing communication across the eons.

If a tetrahedron is circumscribed by a sphere, with its apex anchored at either the North Pole or South Pole, then the three vertices of the base will "touch" the sphere at 19.5° in the hemisphere opposite the polar apex alignment. In addition, the value of e (as in the e/π ratio which is encoded at least *ten times* throughout the Cydonia complex) is 2.718, a near exact match for the ratio of the surface area of a sphere to the surface area of a tetrahedron (2.72).

This whole "tetrahedral" motif was reinforced when they went back to the original Cydonia images. Some of the small mounds Hoagland had noted earlier had the look of tetrahedral pyramids, and the Crater Pyramid, which is involved with one of the key 19.5 measurements, is also tetrahedral. The mounds themselves were also arranged into a couple of sets of equilateral triangles, the 2D base figure for a tetrahedral pyramid.

Later, Dr. Horace Crater, an expert in probabilities and statistics, did a study of the mounds at Cydonia with Dr. Stanley McDaniel. What Crater found was that not only was there a non-random pattern in the distribution of the nearly identical mounds at Cydonia, but that the pattern of distribution was overwhelmingly tetrahedral—and to a factor of 200 million to one against a natural origin.

In 1989, Hoagland and Torun proceeded to publish their results in a new paper titled, appropriately, "The Message of Cydonia." Based on the barrage of personal attacks Hoagland had endured from NASA critics after *Monuments* was published two years before, they assumed that it would be pointless to try to have their paper published in the NASA-controlled "peer reviewed" journals. Instead, they decided to go straight to the people and uploaded the paper to CompuServe, the largest online message board of its time. The paper contained a number of predictions based on their evolving theory of the tetrahedral Message of Cydonia and also the even more radical new idea that within the tetrahedral mathematics was nothing less than an entirely new physics model. Hoagland then found that there was a long-abandoned line of thought among some of the masters of early physics, including James Clerk Maxwell, which included the idea that certain problems in electromagnetics could be solved by the imposition of higher spatial dimensions into the equations. The energies coming from these higher dimensions would then be "reflected" in our lower three dimensional universe through tetrahedral geometric signatures. It was this crucial insight, they decided, that the builders of Cydonia were ultimately trying to impart.

In looking at the planets, which are really just giant spinning spheres themselves, they found that there was a distinct "tetrahedral pattern" to their geology. This expressed itself as energetic outputs — giant disturbances on the gas giant planets, like Jupiter, with its Great Red Spot located at the tetrahedral 19.5 degree latitude, and as great uplifts of landmass on rocky worlds like Mars, where Olympus Mons is located at 19.5 degrees N. Even our own Earth followed this tetrahedral pattern in Hawaii's Mauna Kea volcano, which is larger than Mt. Everest and has been in a constant state of eruption for at least the last million years — and is right around 19.5 degrees S.

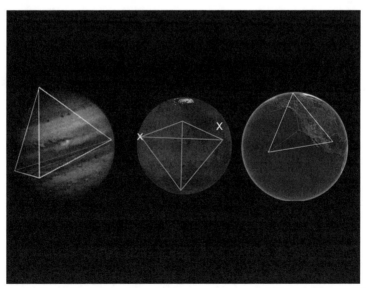

Jupiter's Great Red Spot, Mars' Olympus Mons, and Earth's Mauna Kea volcano all lie at or very near 19.5 degrees latitude on their planetary spheres. Cydonia lies 120° from Olympus Mons and 40° N (marked by second "X")

The mathematics of this "Hyperdimensional physics," as described by theoretical mathematicians like H.S.M. Coxeter, predicted that the behavior of a spinning sphere, such as a planet, would outwell higher dimensional energies at the key tetrahedral latitude—the now-ubiquitous 19.5 degrees. A lesser-known aspect of this model was the further prediction that there would be energetic "inwelling" points in the system as well, and that they would be hexagonal. There had been some confirmation of this idea in images taken of Saturn and the sun. Both sets of images showed hexagonal rings of clouds around the northern poles of the bodies, making the turns at high velocity. No known physical phenomena could account for this behavior (for more on Hyperdimensional physics, see *Dark Mission* and *The Choice*).

The reductionists were quick to attack Hoagland and Torun's ideas about what the Ancient Alien civilization on Mars had been trying to communicate through this monumental mathematics. The critics argued against the validity of the model on one of two basic counts—that the measurements were either inaccurate, or if they were accurate, they did not mean what Hoagland and Torun

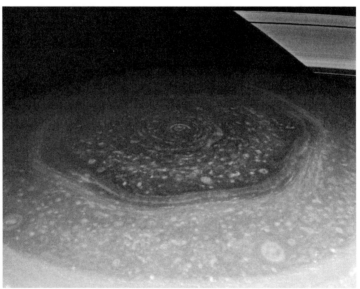

Saturn's north polar hexagon cloud pattern, predicted by Hoagland and Torun's Hyperdimensional physics model.

implied they meant.

Anonymous memos from within NASA in the late 1980s used the same sorts of tactics they had leveled at Dr. Blair years earlier. They argued that Torun's measurements were not reliable because of the amount of error built into the ortho-rectified images. They frequently disputed the measurements themselves, but did not actually bother to try and reproduce them. Dr. Ralph Greenberg, a University of Washington mathematics professor, has more recently taken up this view. Greenberg has written several documents critical of Hoagland and Torun's model, and has also made something of a mini-career for himself accusing Hoagland of lying about his contributions to the idea of life under the oceans of Europa, which has been forcefully refuted by the people who witnessed it.

Dr. Michael Malin of Malin Space Science Systems (who controlled the camera for the then planned *Mars Observer* and the recent *Mars Global Surveyor*) took a slightly different tack, agreeing that the measurements made by Hoagland and Torun "are not wholly in dispute" but arguing that even if the numbers were right, it did not necessarily follow that they meant something significant.

Most of these critiques are typical of the type of reaction you get from scientists when their established paradigms are threatened or when experts in particular fields try to apply the standards they are familiar with to a problem that is outside their experience. The issue of margins of error, especially, is one that (even today) is simply misunderstood, even by experienced mathematicians. Put simply, Greenberg argues—as many have before him—that the margin of error built into the measurements of the Cydonia Complex renders them useless, because they are large enough to make almost "any" mathematical constants and ratios possible. Greenberg, who has become pretty much the point man for attacks on the Cydonia Geometric Relationship Model, also claimed that Hoagland and Torun "selected" the angles they found, implying that they were looking for specific relationships before they ever started. As we established earlier, this is not the case.

For the record, Greenberg also argues that the frequently cited mathematical and astronomical alignments of the Pyramids in Egypt are fallacious, even though no credible Egyptologist doubts them. It is by now well established that the base of the Great Pyramid is a square with right angle corners accurate to 1/20th of a degree. The side faces are all perfect equilateral triangles which align precisely with true north, south, east and west. The length of each side of the base is 365.2422 Hebrew cubits, which is the exact length of the solar year. The slope angle of the sides results in the pyramid having a height of 232.52 cubits. Dividing two times the side length by this height gives a figure of 3.14159. This figure gives the circumference of a circle when multiplied by its diameter. The perimeter of the base of the pyramid is exactly equal to the circumference of a circle with a diameter twice the height of the pyramid itself.

Because of the angle of the slope sides, for every ten feet you ascend on the pyramid, your altitude is raised by nine feet. Multiplying the true altitude of the Great Pyramid by ten to the power of nine, you get 91,840,000, which is the distance from the sun to the earth in miles. If you further divide this distance by the height of the Great Pyramid, you would end up stacking exactly one billion Great Pyramids on top of each other to reach the sun. In addition, the builders also apparently knew the tilt of the earth's axis (23.5°), how to accurately calculate degrees of latitude (which vary as an observer ventures farther from the equator) and the length of the earth's precessionary cycle.

And all of this, according to the brilliant Dr. Greenberg, is just a coincidence. Just an example of "the power of randomness."

Greenberg's arguments are pure reductionism. Forgetting for a moment the sheer unlikelihood of finding consistent and redundant mathematical linkages among a very few objects pre-selected only for their anomalous geology (which Greenberg does not address in any of his arguments) not for their possible mathematical relations to one another, and the fact that only clear and obvious structural points on these objects were used in measurements, Greenberg completely fails to grasp the issue—Hoagland and Torun's measurements are meant to be *nominal*, meaning that they are valid to the closest fit of the methodology employed. They are not saying, "these are the numbers within a loose tolerance range," they are saying flatly "these are the numbers." The tolerances are just what they had to live with pending higher resolution images. Further, having stated that the measurements reflect a specific tetrahedral geometry—not just any set of "significant" mathematical numbers, as Greenberg implies—and that they encode a predictable physics, it becomes very easy to simply test their contextual model vs. his reductionist view. Greenberg seeks to isolate the numbers themselves, and argue only his view of the "power of randomness," rather than simply test the alignments in the greater context of the physics they imply.

Fortunately, "The Message of Cydonia" paper contained predictions that would provide the ideal opportunity to do just that. At that time, NASA's *Voyager 2* probe was approaching Neptune but had yet to image the planet up close. At the end of their paper, Hoagland and Torun put in three specific predictions about what *Voyager 2* would see. They first predicted a storm or disturbance within a few degrees of the tetrahedral 19.5° latitude. Based on their further interpretation of the Hyperdimensional physics they were developing, they also predicted that this disturbance would be in the southern hemisphere of the planet, and that the magnetic dipole polarity of Neptune's magnetic field would be anchored at the Northern pole.

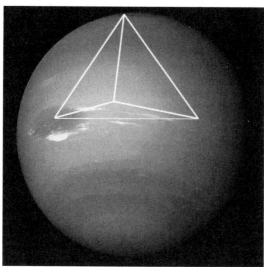

Neptune's "Great Dark Spot," located at 19.5 degrees S, exactly where Hoagland and Torun predicted it would be.

All three predictions—remember, based on the supposedly "fallacious" numbers derived from a supposedly "meaningless" set of alignments of possible ruins on Mars—turned out to be... absolutely correct.

Greenberg and the reductionists then argued: "a single prediction, no matter what it is based on, cannot be relied upon as proof of anything." This tactic, combining the predictions into a single one instead of three, is a common means of dismissing the frequency of Hoagland and Torun's successes. As Harvard astronomer Halton Arp put it in his excellent book *Seeing Red*, "The game here is to lump all the previous observations into one 'hypothesis' and then claim there is no second, confirming observation."

There is, flatly, no way that Hoagland and Torun could use a set of "meaningless" or "fallacious" data to make three such accurate predictions about features on a planet the human race had never seen up close before. These features have no explanation in the conventional models, at least as far as providing a mechanism for the storm, the location of the storm, and its relationship to the magnetic pole of the planet. In other words, there is no way they could have just gotten lucky by using established models of the solar system. Their predictions come solely from the Cydonia Geometric Alignment model. This is not only a ringing endorsement of the validity of both the measurements and the physics model deduced from them, but also a harsh indictment of the methods and motives of both Greenberg and Malin (Greenberg at one point challenged Hoagland to a "debate" on the mathematics of Cydonia, but only if he could exclude Crater's tetrahedral mound data, which he acknowledged he could not explain away).

By the early 1990s, all that was needed to resolve the arguments were higher resolution images of Cydonia, the Face and the other objects in question. Since NASA had a new probe with a better camera named *Mars Observer* that was scheduled to arrive in 1993, it seemed that the debate would soon be settled once and for all.

But then, as they say "NASA happened..."

126

(Endnotes)
1 Hoagland, Richard C., *The Monuments of Mars, a City on the Edge of Forever,* North Atlantic Books, ISBN-13: 978-1583940549
2 http://www.imdb.com/title/tt0957961/combined
3 http://users.starpower.net/etorun/pyramid/geomorphology.html

CHAPTER 5
THE POLITICS OF MARS

Unfortunately, as the decade of the 90s turned, the search for life on Mars had taken a decidedly political turn. This had actually all started back in 1976 and included not just the brief controversy over the Face and NASA's lies about it, but with the actual microbial life test results from the *Viking* landers themselves.

As we have seen, the prevailing NASA view of Mars (to which most planetary scientists in and outside NASA have quietly acquiesced) was that it was a "dry, dead desert planet" with atmosphere so thin that not even water could stay in a liquid state on its surface for more than few seconds. By the time the two *Vikings* had arrived on Mars in summer 1976, the only real hope was that primitive microbial life might be found there, functioning at a low level, but surviving. The two landers had several experiments designed to test for just that possibility. Advocates of the idea had always pointed to one key clue that had been observed since the days of Schiaparelli and Lowell: the so-called "wave of darkening" that was observed to sweep from each hemisphere's pole toward the equator in their respective spring seasons. This darkening wave rolls across the planet at a rate of thirty-five miles per day, and was once attributed to melting polar ice caps releasing water into the atmosphere and "awakening" the planet's simple plant life. This was disputed at the time of *Viking* because it was later found that the southern polar cap was entirely carbon dioxide (dry) ice. However, *Mars Odyssey's* new 2001 observations found that there were vast quantities of water ice all over the planet that could theoretically fuel these darkening waves. While the reality of the wave of darkening is still disputed by some, no one argues that certain patches on Mars do darken in the spring and summer.

The view of Mars as stark and sere prevailed in the mid 1970s, when the two *Viking* landers were sent to test the soil for signs of microbial life. What most people do not remember is that the lander tests for life both came back positive.

The *Viking* landers actually carried four different experiments that were designed to look for signs of life. Of these, the most anxiously awaited and promising was the Labeled Release experiment, or LR. The LR experiment took a sample of Martian soil and introduced a drop of water loaded with seven different nutrients into it. These nutrients were "labeled" radioactive Carbon-14. The instrument then scanned for the presence of radioactive 14-CO_2 gas. The release of this gas would be an indication that organic, living microbes had actually metabolized the nutrients and that would

Trench dug by *Viking 1* soil sampler scoop on July 28th, 1976.

constitute proof that life was present in the Martian soil.

And that is exactly what the *Viking 1* LR experiment found.

In fact, both *Viking* landers got the same result despite being done with samples taken under very different conditions. The *Viking 1* sample came from the surface, which had been exposed to sunlight (the Martian atmosphere has no ozone layer which exposes any life to damaging ultraviolet rays), while the *Viking 2* sample had been taken from beneath a rock; both tests came back positive.

NASA, however, quickly moved to suppress this news and present an alternative view—that the results were just a mistake, a chemical reaction and not proof of life on Mars. They argued that the other three tests had come back negative, and therefore there must have been some unforeseen chemical process involved. However in 2006 researcher Rafael Navarro demonstrated in a paper that the three other *Viking* biological experiments likely lacked the sensitivity to detect trace amounts of organic compounds—in other words, they were flawed from the beginning.[1]

The NASA view that it was all just a funky chemical reaction has also been hotly disputed by the man who designed the tests in the first place, Dr. Gil Levin of Biospherics Inc. laboratories. Dr. Levin has always insisted that his instruments' results were positive for life, and not a result of a mere "chemical interaction." His case was bolstered in 1996 when NASA announced the discovery of microfossils in a meteorite from Mars. Obviously, if there were once micro-organisms living on Mars, there was no reason that they could not be present on Mars today. The only remaining argument against that conclusion was the supposed absence of a "biologically kind" environment, i.e. liquid water.

Levin himself had argued for some time that this was not really an issue. He presented a paper describing the circumstances under which water could remain in a liquid state on Mars.[2] He pointed out that the NASA view of Mars as unable to support water at the surface was based on a faulty assumption—that the water was evenly distributed throughout Mars' atmosphere, rather than in "the lower one to three km," as confirmed by the later *Pathfinder* probe.

So, the reality was that there was plenty of evidence that

Mars was not only capable of harboring life, but that NASA had in fact already proven that to be the case, as far back as 1976. Levin's co-worker on the Labeled Release experiment, *Viking* scientist Patricia Ann Straat, commented to *Discovery News* that "Our (LR) experiment was a definite positive response for life, but a lot of people have claimed that it was a false positive for a variety of reasons."[3] NASA's determined efforts to suppress such a conclusion would seem to fly in the face of the agency's publicly stated mandate. But taken in the broader context of how the other *Viking* data had been handled, it fits with what seems to a political, rather than a scientific, agenda.

THE TRUE COLORS OF MARS

This odd political agenda of suppressing the truth about Mars went beyond just the Face and Cydonia and reached all the way from NASA headquarters to the halls of the Jet Propulsion Laboratory in Pasadena, California. For some reason, NASA has an aversion to showing the world the "true colors of Mars."

The first hint of a political agenda on the issue came just after the first *Viking 1* color images of Mars were released to the world. Within a few hours of that historic publication another hurriedly revised version was suddenly produced, supposedly correcting the initial "color engineering problems" in the first image. But, as recounted his book *Mars: The Living Planet*, science writer Barry DiGregorio showed that Dr. Levin and his son Ron (now a physicist at MIT) remembered things being quite different.
[See Color Plate #1]

"At about 2:00 p.m. PDT, the first color image from the surface of another planet, Mars, began to emerge on the JPL color video monitors located in many of the surrounding buildings, specifically set up for JPL employees and media personnel to view the *Viking* images. Gil and Ron Levin sat in the main control room where dozens of video monitors and anxious technicians waited to see this historic first color picture. As the image developed on the monitors, the crowd of scientists, technicians and media reacted enthusiastically to a scene that would be absolutely unforgettable—

Mars in color. The image showed an Arizona-like landscape: blue sky, brownish-red desert soil and gray rocks with green splotches.

"Gil Levin commented to Patricia Straat (his co-investigator) and his son Ron, 'Look at that image! It looks like Arizona.'

"Two hours after the first color image appeared on the monitors, a technician abruptly changed the image from the light-blue sky and Arizona-like landscape to a uniform orange-red sky and landscape. Ron Levin looked in disbelief as the technician went from monitor to monitor making the change. Minutes later, Ron followed him, resetting the colors to their original appearance. Levin and Straat were interrupted when they heard someone being chastised. It was Ron Levin being chewed out by the *Viking* project director himself, James S. Martin, Jr. Gil Levin immediately inquired as to what was going on. Martin had caught Ron changing all the color monitors back to their original settings. He warned Ron that if he tried something like that again, he'd be thrown out of JPL for good. The director then asked a TRW engineer assisting the Biology team, Ron Gilje, to follow Ron Levin around to every color monitor and change it back to the red landscape.

"What Gil Levin, Ron and Patricia Straat did not know (even to this writing) is that the order to change the colors came directly from the NASA administrator himself, Dr. James Fletcher. Months later, Gil Levin sought out the JPL *Viking* imaging team technician who actually made the changes and asked why it was done. The technician responded that he had instructions from the *Viking* imaging team that the Mars sky and landscape should be red and went around to all the monitors, "tweaking" them to make it so. Gil Levin said, "The new settings showed the American flag (painted on the Landers) as having purple stripes. The technician said that the Mars atmosphere made the flag appear that way."

One of the basic questions that should have been asked involves the physics behind JPL's abrupt color alterations. As Dr. Gil Levin phrased it in DiGregorio's book:

"If atmospheric dust were scattering red light and not blue, the sky would appear red, but since the red would be at least partially removed by the time the light hit the surface, its reflection from the surface would make the surface appear more blue than red. There would be less red light left to reflect. And what about the sharp shadows of the rocks in the black and white images yesterday? If significant scattering of the light on Mars occurred, the sharp shadows in those images would not be present, or at best, would appear fuzzy because of diffusion by the scattering."

Levin was describing the well-known phenomenon of "Raleigh scattering" whereby the similar-sized molecules of all planetary atmospheres (be it the primary nitrogen of Earth, the carbon dioxide atmosphere of Mars, or even the predominantly hydrogen atmospheres of Jupiter and Saturn) all produce blue skies when sunlight passes through them. If you examine the long Martian photographic record—which encompasses hundreds of thousands of images acquired by dozens of observatories even before the Space Age dawned—you can see blatant evidence that Levin is right and JPL is wrong.

So we have a contradiction. If Mars has a blue sky and looks remarkably like "Arizona" from the ground, why does NASA keep making the images from the surface appear Technicolor red? What possible agenda could it serve to deceive the public about the true colors of Mars? An answer might be found if we go back to *Mars: The Living Planet*.

It turns out that DiGregorio's statement that the NASA administrator was behind the monitor changing incident was based on a confirmation of this from an official source—former JPL public affairs officer Jurrie J. Van der Woude—and it had an even stranger and somewhat sinister angle. In a letter to DiGregorio (also reproduced in *Mars: The Living Planet*), Van der Woude wrote:

"Both Ron Wichelman [of JPL's Image Processing laboratory (IPL)] and I were responsible for the color quality control of the *Viking Lander* photographs, and Dr. Thomas Mutch, the *Viking* Imaging Team leader, told us that he got a call from the NASA Administrator asking that we destroy the Mars blue sky negative created from the original digital data."

This bizarre sequence of events raises many disturbing questions. For instance, why was the administrator of NASA so determined to conceal the "true" colors of Mars from the American people and the world in 1976? Why would he order the head of the Viking Imaging Team to literally eliminate an important piece of historical evidence from the official mission archive—the original "blue-sky negative"—if the initial release was only an honest technical mistake? Wouldn't that record be an important part of the ultimate, triumphant story of NASA scientists correcting initial scientific errors, in their continued exploration of the frontier and alien environment of another world? And why would a young teenager (the son of one of the key investigators on the *Viking* mission, no less) be threatened with expulsion by the director of the project for simply tweaking a couple of color monitors around the lab?

In truth, none of Ron Levin's story (or Van der Woude's significant confirmation), makes any scientific sense unless certain individuals at the highest levels in NASA felt compelled—for some arcane reason—to hide at all costs the visible appearance of the actual Martian surface.

[See Color Plate #2]

Beyond that, there is an even bigger "biological problem" for the conventional NASA view of the true colors and environment of Mars. Levin suggested that there were other hues on Mars than just dull browns and reds. This was verified by members of the *Viking* imaging team, who confirmed there were blue and green patches on rocks that changed seasonally. The only rational explanation for these "changing patches" on the rocks, shifting color with the rising and dropping seasonal temperatures and atmospheric availability of water is biological entities, like simple plants or lichens, reacting to changing biospheric conditions.

This pattern has continued through mission after mission, as NASA has altered the color of images of the surface of Mars to make it appear as a red and alien landscape rather than the blue-skied, Arizona-like desert that it really is. But, as the new decade of the 90s dawned, NASA had much bigger problems to deal with. They were going back to Mars, and under incredible pressure to take more pictures of objects they had denied were even there.

Artist's rendering of the ill-fated Mars Observer spacecraft.

MARS OBSERVER

Mars Observer was announced in the late 1980s as the next generation follow-up to *Viking*. The mission would be the first new unmanned reconnaissance of Mars in almost 20 years, with a host of vastly improved scientific instruments. However, initial specs for the spacecraft were highly disappointing to anyone seeking a resolution to the Cydonia issue, since the mission was not originally designed to even include a camera. Eventually, the mission planners came to their senses, and it was decided fairly late in the game to include a one-meter-per-pixel resolution grayscale camera. That, however, was where the problems actually began.

The man who would build, point and control the camera was a former JPL employee named Dr. Michael Malin. Among his other interesting affiliations, he had once been part of a project to analyze the purported UFO photographs of infamous "contactee" Billy Meier. In that capacity, Malin, then an associate professor at

Arizona State University, had concluded that Meier's controversial photographs were not fakes.

Malin chose to reserve judgment on the more spectacular aspects of Meier's story, but this early foray into such arcane territory showed that he was at least willing to consider unusual or even bizarre claims like Meier's. What was in question however, was just what his position was on re-imaging Cydonia and the Face.

Malin quickly asserted that he had no interest in even testing the Cydonia hypotheses by targeting the formations with his new camera. In fact, he stated his outright opposition to making even a minimal effort to re-photograph Cydonia on numerous occasions. Because the camera was a "nadir pointing" device, meaning that it could not swivel or aim at specific targets without the entire spacecraft being repositioned (and hence using valuable fuel), Malin argued that at best he might get "one or two" random opportunities to target a specific object like the Face or D&M pyramid during the regular science mission. However, as the specs evolved, *Mars Observer* soon became a much more capable mission, with additional fuel added to the mission plan to enable an extension of the original two-year science acquisition phase of the project.

Hoagland and Dr. Stanley McDaniel began to dig into Malin's contentions, and quickly discovered that Malin's claim of at best "one or two" opportunities to target the Face was greatly understated. After consulting with mission planners at JPL and reviewing the technical specs, they found that there would be more on the order of 40+ chances to target the Face during the regular two-year science phase. So why would Dr. Malin—if he was honest—underestimate the imaging opportunities by a factor of twenty? Hoagland and McDaniel smelled a rat, and they decided to try an end run.

Hoagland and the other researchers then began to lobby NASA and Congress to target the formations, only to make an extremely unpleasant discovery. Neither NASA nor Congress had anything to say about where Michael Malin pointed his Mars Orbiter Camera.

In an unprecedented move, NASA had decided to sell the rights to all of the data collected by the *Observer* to Malin himself, in an exclusive arrangement that gave Malin godlike powers over when, or even if, he decided to release any data the camera collected. This private contractor arrangement not only neatly absolved NASA from any responsibility as to what was photographed with an instrument and mission paid for by the taxpayers of the United States, but it gave Malin the right to embargo data for up to half a year, if he so chose.

This marked the first time in NASA history that data returning from an unmanned space probe would not be seen "live," as it had all throughout the preceding 30 plus years of *Mariner*, *Lunar Orbiter*, *Surveyor*, *Apollo*, *Viking* and *Voyager* missions. The logic of the arrangement was tenuous, at best. NASA claimed that in order to assure that private contractors would bid on future space projects like *Mars Observer*, they had to guarantee an "exclusive rights period" to the private contractors/scientists, so that they could write the first scientific papers from the data collected "without unfair competition from other, non-project scientists."

Of course, it was not required in any way to grant Malin the right to withhold some or all of the data completely, which he could, under a clause that gave him the right to delete "artifacts" from any or all of the images. In essence, Malin could release a blacked-out image, and then simply claim the image had been filled with artifacts. "Artifacts" in image processing are flaws or blemishes appearing on the image; they can arise during processing but are often the result of dirt in the optical path or energy rays affecting the sensors during imaging. The clause also meant that for a period of up to six months, Malin could do literally anything at all to the images, and no one—not even NASA—would be the wiser.

Malin even moved his entire private company, Malin Space Science Systems (which held the actual *Mars Observer* camera contract) away from ASU in Arizona and JPL in California to San Diego. This effectively insulated Malin from the Mars planetary science community. Visitors—other scientists within the community, or even co-investigators with Malin on *Mars Observer*—were quite unlikely to "drop in" unannounced if everyone had to drive four or five hours from JPL just to get to Malin's offices. And, when they did get there, if they didn't get directions beforehand, they'd be out of luck. For some reason, Malin's company was never publicly listed on the shopping mall marquee where his offices were actually located.

Curiously, however, the move did put him right across the street from one of the world's largest "supercomputer" facilities, where he could literally hand-carry digital imaging tapes back and forth.

To Hoagland and the other independent researchers, this was an untenable situation. It was anathema to Hoagland that a publicly funded program could be subject to such an obvious sell-out of the public's right to know, and their faith in the integrity of the data. Instead, the total control was in the hands of a man who had expressed outright hostility to the idea of even testing the Cydonia hypothesis. So the whole issue was subject to Malin—without oversight of any kind—having the scientific integrity not to alter or withhold data that might make him look like a fool.

By 1992, with the September launch of *Mars Observer* rapidly approaching, Dr. McDaniel entered the fray. Using various political and academic contacts, he began to put pressure on NASA and JPL from several directions, forcing them to address, on the record, just why they were not able to target Cydonia or the Face specifically. NASA responded with various contradictory, if not disingenuous (McDaniel's words) arguments, including those by Dr. Malin. At each and every turn, McDaniel and Hoagland shot down the arguments, even finally getting NASA Headquarters Public Affairs' spokesman, Don Savage, to officially admit (in a Headquarters letter) that the infamous "disconfirming photos" of the Face never existed.

Mars Observer was a troubled mission almost from the very beginning. Besides the various political controversies swirling around the Cydonia question, it had a series of technical mishaps that made even casual observers wonder if the mission was cursed, or if somebody just didn't want it to succeed. Even the mission's Project Office described *Mars Observer*'s journey to the Red Planet as "traumatic."

In late August 1992, during a routine inspection of the

spacecraft on the launch pad, NASA technicians discovered severe contamination, inexplicably inside the protective shroud, consisting of "metal filings, paint chips and assorted trash." NASA publicly speculated that the damage was done when the spacecraft had been hastily unplugged from an outside air-conditioner and the payload shroud hermetically sealed, a measure actually designed to protect it from the imminent effects of Hurricane Andrew. But the agency never actually cited a specific cause for the contamination from its (brief) investigation. With an immovable launch window looming just weeks away, the orbiter payload was hurriedly removed from the pad and taken back to the payload integration clean room—for disassembly, inspection and possible "aggressive cleaning."

It was there that Program technicians made a second, even more disturbing discovery.

According to *Mars Observer* Project Manager David Evans, during the inspection process NASA discovered the presence of an unspecified "foreign substance" inside the spacecraft's (Malin's) camera assembly that would have made the resultant images blurred and virtually worthless for resolving the Cydonia issue. According to Evans, since the camera was a sealed assembly, the mysterious contamination could only have been introduced into the camera in disassembly and check out *after it came assembled from Malin's facility*—in the JPL clean room itself.

How such a basic "mistake" could have been made, given the nearly $1 billion price tag of the mission, is hard to fathom. Checking the cleanliness of the camera optics is invariably the top priority for a mission that has a visible light camera as its primary scientific instrument. Had this bizarre Vaseline-like smearing of the lens not been accidentally discovered at the Cape, *Mars Observer* would have been an embarrassment on the scale of the original Hubble Telescope debacle. Fortunately, NASA engineers at Cape Canaveral were able to clean the spacecraft and get it back to the pad in record time for its September 25 launch.

Meanwhile, NASA management was no longer simply insisting that the terms of Malin's private contract with the agency gave him the "right" to target or ignore Cydonia at his whim (as well as embargo the images and legally remove "artifacts" from the data); Program Scientist Bevan French was additionally defending the notion that the Face and other objects were "too small" to be effectively targeted by the Malin camera in the first place. This was despite the fact that there was a defined mission objective to target the sites of the two previous *Viking* landers which, as opposed to the mile-wide Face, were each less than 15 feet wide.

As the launch date arrived, the political pressure was reaching a fevered level; Hoagland was live on CNN, reminding viewers of all this strange history even as the spacecraft lifted off. Fortunately, the actual launch itself seemed to go off without a hitch. Then, something truly bizarre happened: all contact was lost with *Mars Observer* and its still-attached second stage rocket, for almost 90 minutes.

Just 24 minutes into the mission, with the spacecraft set to fire a second-stage rocket after separating from its first stage Titan booster, all radio and telemetry went dead. Aircraft over the Indian Ocean reported seeing a brilliant red-orange flash—possibly the second stage firing, possibly the spacecraft exploding—coinciding with the timing of the critical rocket firing. Given that the spacecraft had gone inexplicably silent, flight controllers assumed the worst. Imagine their relief when a little over an hour later, *Mars Observer* just as suddenly and inexplicably reappeared, apparently none the worse for wear.

So what exactly had happened during those lost 85 minutes?

It's impossible to know for sure, but on two subsequent attempts to retrieve the onboard telemetry recorded during the "missing time" event, there was absolutely nothing to be heard. Then, on a third attempt—several days later – suddenly, a completely normal data stream appeared. There was only one problem: the first two attempts had received a carrier signal and "timing code," indicating that a recording was made, but the tape simply contained no data. How did the data from the missing-time episode suddenly find its way onto a tape that had been blank only days before? It was as if someone had erased the actual recording, then subsequently uploaded a manufactured "nominal" data stream a few days later.

The Deep Space Network (DSN) engineers were insistent that they hadn't simply missed something the first two times around. "There was no data on that tape the first two times!" JPL's Deep Space Network manager angrily declared.

The news media, of course, had little knowledge or understanding of just how impossible the whole situation was, and quickly dropped the issue. It did, however, become considerably more relevant eleven months later.

By that time, after a relatively quiet trip toward the Red Planet, *Mars Observer* was nearing its goal and the debate over Cydonia was once again gaining steam. News stories mentioned Cydonia as a matter of course. Then, just weeks before *Mars Observer's* scheduled orbital insertion burn and the delivery of McDaniel's report to both Congress and NASA, the agency suddenly decided to change plans. NASA indicated a willingness to reconsider not only its position on the data embargo and the lack of live televised images from the orbiter, but also announced that they were considering a radical new science plan.

Because the first few weeks of the planned mapping orbit period would occur during a solar conjunction and just before the beginning of dust storm season on Mars, there was a chance it could be months before any pictures of Mars were returned at all, much less targeted images of Cydonia. NASA's solution was to try a "power-in" maneuver that would place the spacecraft in a mapping orbit some 21 days early. However, in other documents and letters to Congress, NASA inexplicably added almost as many days to the "check out" and calibration phase upon reaching this science mapping orbit, meaning that no useful images of the planet could be expected until after the conjunction, at the least.

To Hoagland and McDaniel, the sudden lengthening of the unnecessary "calibration" phase was an obvious ruse. If JPL was going to take extra time to "calibrate" the instruments, effectively negating the advantage of the power-in maneuver, why bother "powering in" at all? The answer seemed simple: by powering in, NASA could buy themselves 21 priceless days to secretly examine whatever Martian real estate they wished (obviously Cydonia) without any public or media pressure to release the data they were gathering.

Any and all of the images acquired in this time period could be "officially" denied, since the spacecraft was simply being "calibrated" and not really gathering science quality data at all.

Predictably, Hoagland and McDaniel raised a stink, and NASA suddenly found itself under additional pressure from various sources to provide live images of Cydonia. Hoagland upped the ante by scheduling a press conference for the day that *Mars Observer* was scheduled to achieve orbit around the Red Planet. The briefing would be held at the National Press Club in Washington, D.C., and would be attended by many of the principals involved in the independent investigation, including Dr. Mark Carlotto, Dr. Tom Van Flandern, Dr. David Webb and architect Robert Fiertek.

And then, four days before *Mars Observer* was scheduled to make its orbital burn and commence operations, McDaniel delivered his final report simultaneously to NASA, Congress, the White House and the media. *Mars Observer* mission director Bevan French got a personal, hand-delivered copy. The following Sunday, August 22, 1993, French was scheduled to debate Hoagland on national TV, on ABC's *Good Morning America.*

Hoagland destroyed French in the open forum. Having been given an astonishing six minutes, more than twice the usual time allotted for such segments, Hoagland used the opportunity to

"Good Morning America" -- 1993

The Hoagland/French debate *Good Morning America*, August 22nd. 1993 bludgeon French's weak and sometimes contradictory arguments. Forced to defend an indefensible position—that NASA should willfully allow one man to have godlike powers over data paid for by the American taxpayers—French wilted under the pressure. The final insult came at the end, when the exasperated host, Bill Ritter, finally just confronted French point blank. "Dr. French, why don't you just take the pictures, immediately release them and then prove these guys wrong?" French, unsurprisingly, had no real answer.

Then, at exactly 11:00 AM Eastern Time, just moments after Hoagland had creamed French on national television, AP science reporter Lee Siegel got a call from a JPL spokesman. The NASA rep informed him that *Mars Observer* had simply disappeared, some 14 hours earlier. The timing of this announcement, just moments after the *Mars Observer* Program Scientist had badly lost a very public nationally televised debate with the leader of a highly critical agency as an opponent, seemed a bit too coincidental. Why hadn't French simply admitted that the *Mars Observer* was lost at the top of the segment? It is inconceivable that he, the Program Scientist, didn't know for over 14 hours that "his" spacecraft had been lost. French could have saved himself a lot of heat and needless embarrassment by simply announcing on *Good Morning America* that the Mars Observer was in trouble. This would have neatly changed the subject of the segment, and shifted any discussion of Cydonia and artifacts to the back burner.

In hindsight, it isn't difficult to figure out what actually happened. After other high NASA officials (and their bosses) watched French's lame Cydonia spin control fail miserably—and on live television—NASA went to Plan B. They either pulled the plug on the Mission outright out of fear of what uncensored images of Cydonia would reveal, or NASA simply took the entire Mission "black." The extraordinary scrutiny the agency was under at the time would have made it nearly impossible to conduct a survey of Cydonia in secret. The most viable solution, when faced with this sort of pressure, was to either scrap the program, or come up with a way to conduct the preliminary reconnaissance in secret – hidden not only from the general public and the press, but from its own

144

"honest" employees at JPL as well. As it happened, NASA pulled off exactly that scenario, under rather suspect circumstances. With *Mars Observer* officially "lost," they could conduct a highly detailed survey that could tell them either how to take future "public" images to ensure minimum political impact, or even how to whitewash the images believably.

An official commission was formed to determine what had caused the spacecraft to cease operations. Unfortunately, the investigation was doomed from day one for one simple reason: there was no engineering telemetry to analyze.

NASA, in another unprecedented move, had inexplicably ordered *Mars Observer* to shut off its primary data stream prior to executing a key pre-orbital burn. Resultantly, there was no data at all from the spacecraft's final few nanoseconds of existence, if indeed it had been lost. This is crucial, since even if a chemical fuel explosion had taken place, it would obviously travel much slower than a speed of light radio signal, and the spacecraft's destruction sequence could have been recorded. Such a recording could have been used to reconstruct those final moments in detail and make an educated determination as to exactly what had gone wrong. Instead, because NASA had violated the first rule of space travel—you

The only photograph *Mars Observer* ever took of the Red Planet

never turn off the radio—no cause for the probe's loss was ever satisfactorily determined. All the commission could do was take a wild guess at what might have happened. In the end, *Mars Observer* only ever took one "official" photograph of the Red Planet.

All of this may seem cloak-and-daggerish, but the facts are there to support the idea that something very fishy was going on with the *Mars Observer* from the beginning. From the inexplicable pre-launch "sabotage," to the mysterious loss of signal for over an hour (when an alternative set of instructions could have been uploaded to the spacecraft unbeknownst to the regular spacecraft launch crew or flight controllers), to the ill-conceived "power-in" deception, followed by the bizarre behavior over the loss of the spacecraft (withheld by the project head until minutes after he had lost a crucial debate with Hoagland), nothing seemed normal about this mission.

In the end, the *Mars Observer* debacle was just an interim step in the politicization of the Ancient Aliens on Mars question. The dirty politics from NASA in fact, was about to get a lot worse.

(Endnotes)

1 Biemann, Navarro-Gonzalez; Navarro KF; de la Rosa J; Iniguez E; Molina P; Miranda LD; Morales P; Cienfuegos E; Coll P; Raulin F; et al. (2006). "On the ability of the Viking gas chromatograph–mass spectrometer to detect organic matter". Proc Natl Acad of Science (PNAS) 104 (25): 16089–16094. Bibcode: 2007PNAS..10410310B. doi:10.1073/pnas.0703732104. PMC 1965509. PMID 17548829

2 Bianciardi, Giorgio; Miller, Joseph D.; Straat, Patricia Ann; Levin, Gilbert V. (March 2012). "Complexity Analysis of the Viking Labeled Release Experiments". IJASS 13 (1): 14–26. Bibcode:2012IJASS..13...14B. doi:10.5139/ IJASS.2012.13.1.14. Retrieved 15 April 2012.

3 "Viking 2 Likely Came Close to Finding H2O." Irene Klotz, *Discovery News*, Sept. 28, 2009

CHAPTER 6
MARS PATHFINDER

"Still, we cannot deny that the act of placing a tetrahedral object on Mars at latitude 19.5 contains all the necessary numbers and symbolism to qualify as a "message received" signal in response to the geometry of Cydonia. Moreover, such a game of mathematics and symbolism is precisely what we would expect if NASA were being influenced by the type of occult conspiracy that Hoagland, for one, is always trying to espouse."

—Graham Hancock, *The Mars Mystery*

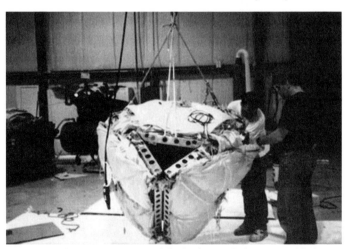

The tetrahedral-shaped *Mars Pathfinder* lander being assembled at JPL.

Shortly after the mysterious disappearance of *Mars Observer*, NASA announced plans for two new even more ambitious missions to the Red Planet. The first of these was to be *Mars Pathfinder*, a mission designed to put a small lander equipped with a rover on the surface of Mars in the summer of 1997. The second mission, *Mars Global Surveyor*, was intended as a replacement for *Mars Observer* and would arrive at Mars in September, 1997.

While the lander would have cameras theoretically superior to the *Viking* landers for imaging the surrounding landscape, the big selling point of the mission was the tiny rover, named *Sojourner*, which would exit the lander and drive around the nearby Martian surface, taking various instrument readings on the rocks it encountered. However, unlike *Viking*, no instruments for testing for microbial life would be included because *Pathfinder* was considered a rover technology demonstration mission. Or at least that was the excuse.

Right away, there were several things about the *Pathfinder* mission that made it interesting to the researchers studying the *Viking* era Cydonia images. The first was that the lander itself, protected by an innovative set of airbags that allowed it to be dropped from low altitude and bounce to a halt on the Martian surface, was shaped like a tetrahedron. Given the overwhelming "tetrahedral" nature of Hoagland and Torun's decoded "Message of Cydonia," they found this in and of itself quite fascinating. But it got even more bizarre when the landing site was announced.

NASA had decided to put the lander down inside a landing ellipse in an ancient flood plain in Mars's northern hemisphere called "Ares Vallis" centered around the coordinates 19.5° N x 33.3° W. Obviously, the 19.5° number had been noted numerous times in the alignments cited by Hoagland and Torun, and the number 33.3° was also a "tetrahedral number," which fell out of the

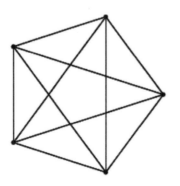 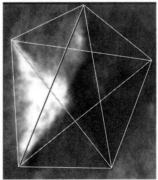

Pentatope (L) and the D&M pyramid on Mars (R).

mathematics of a circumscribed tetrahedron. The numbers "3-3-3" were also significant in the higher mathematics of a Pentatope, an ideal geometric form which represented a pentagon that existed in more than three dimensions (see *Dark Mission*). The Pentatope also bore more than a passing resemblance to Torun's "ideal geometric form" of the D&M pyramid at Cydonia.

When *Mars Pathfinder* finally bounced to a halt after its meteor-like decent to the Martian surface on July 4, 1997, the actual landing site was well within that target ellipse at 19.13 N x 33.22 W. While few people made note of this or the "tetrahedral" connection, the fact is the *Pathfinder* mission and its landing site could hardly have been more clearly tetrahedral. *Pathfinder's* unique tetrahedral design, coupled with the totally repetitive tetrahedral geometry of the landing site, indicated NASA was paying close attention to the Cydonia research that had been done in the previous decade. It was almost as if NASA, perhaps just like the Navy during the 1922 Mars opposition, and despite its protestations to the contrary, had taken the "Message of Cydonia" to heart and wanted to send a symbolic message back—but to *whom*?

Screen capture of live CNN broadcast, July 4th, 1997.

149

Within hours after its successful touchdown, *Pathfinder*, according to a pre-loaded on-board computer program, began to transmit image after image of the scene around the landing site. In fact, there were far more images broadcast on live TV than had been originally advertised by NASA. Contrary to JPL's highly conservative pre-announcements, the images were not just of the "looking down at the edges of the spacecraft solar panels and the airbags" variety—but were of the entire Martian panorama from the edge of the spacecraft to the distant horizon. Watching these initial panoramas it at first seemed that we were simply looking at another boring Martian desert filled with rocks, just like the scenes relayed from the previous *Viking* lander missions some 20 years before. But it was in those first, unplanned and totally uncensored images, that several startling features—clearly *not* belonging on a lifeless desert world—could unquestionably be seen.

Pyramidal "capstone" seen in first-day TV broadcasts of the *Mars Pathfinder* landing site.

Worldwide viewers, watching the NASA unmanned mission live on CNN, began calling in about the strangely geometric rocks that they were seeing in the sudden flood of images. The number of calls was so significant that CNN's science anchor, the late John Holliman, eventually felt compelled to ask one of the *Pathfinder* scientists about it on the air.[1] Geologist Ron Nicks, who had more than 30 years of field experience in the desert southwest, also began looking, and started independently seeing interesting inconsistencies about the landing site as the panoramas swept across the screen. Nicks saw multiple, strangely geometric anomalies around the lander showing up on CNN's live television broadcasts.

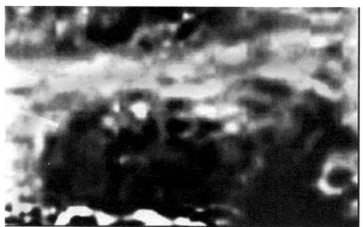

The "Gyro" captured on live CNN broadcast July 4th, 1997.

Nicks and Hoagland, taking careful notes and taping everything from two separate locations, quickly realized that what they were seeing was not the expected wind or water eroded rocks of Ares Vallis, but a debris field filled with a variety of apparently corroded, *manufactured* metal objects. Nicks pointed out what looked to him like canisters with opposable handles, oddly geometric "pointy stuff"—and even a couple of deformed but recognizable tracked vehicles about the same size as the *Sojourner* rover itself. A totally separate research effort—the Near Pathfinder Anomaly Analysis Group (NPAA)—also sprang up, triggered by the same anomalous images appearing live on television, and pointed out a number of artificial looking objects in the near-field area of the lander.

However, after the initial windfall of those first new Mars surface images seen by tens of millions live around the world, the crucial digital post-landing analysis of these anomalies was

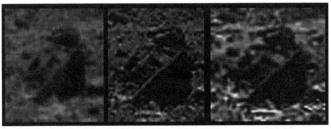

Three views of the "hydrant," an eroded angular mechanism
some distance from the *Mars Pathfinder* lander (NPAA).

immediately frustrated by the maddening lack of clear *Pathfinder*
web images. Even though over 20 years had elapsed between
Viking's mechanical facsimile-type surface cameras in 1976 and
Pathfinder's state-of-the-art CCD imaging technology of 1997, the
images published by NASA were inexplicably worse than those
from *Viking*. Unusual discrepancies between the video record live
on CNN and the NASA web postings were found repeatedly, with
the end result that the TV images were always better. [2] This should
have been exactly the opposite.

The simple truth was that the processed internet digital files
should have been much sharper than the original raw video feed
coming from *Pathfinder* that first afternoon, especially after the raw
feed was rebroadcast over standard definition television's limited
resolution of 640x480. Instead, JPL's web versions were full of
astonishingly amateurish compression artifacts, assembly errors
and horrific color registration problems. JPL even adjusted the color

The "Gyro," a cased metal object with a flywheel-like internal mechanism
lying just beyond the *Mars Pathfinder* lander. NASA image 80881_full.jpg.

of the Martian sky to appear the traditional "*Viking* pink" instead of the expected natural, Earth-like blue. [see Color Page 3]

One of the most interesting objects that appeared in the TV scans and the digital imagery was nicknamed the "Gyro" and sat just beyond the edge of the lander in NASA image 80881_full. jpg. In this image, it can be seen fairly clearly that the Gyro is a cased object with what appears to be a flywheel-like motor inside it. Despite the poor resolution and artifacts present in the image, the regularly spaced flywheel segments can be clearly made out emanating from a central cap. It looks, quite frankly, like a jet engine turbine intake or even a gyroscopic inertial measuring unit. Even though it's probably not either of those, it is unquestionably some kind of cased mechanical device.

Comparison of the Gyro and a jet engine intake.

Another image of the Gyro taken a few days later on 6 July, 1997 showed it again, but this time it was unviewable because NASA simply wallpapered over it.

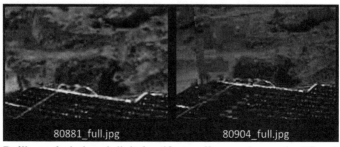

Deliberately induced digital artifact wallpapered over the "Gyro" in NASA image 80904_full.jpg (R). Earlier image (L).

Deliberately mis-registered color image of the Gyro (shown in grayscale) from NASA image 80809_full.jpg.

Later color images of the Gyro were so badly mis-registered that they were utterly useless for comparison. Somehow the genius image enhancement experts at NASA had managed to overlay the red, green and blue filter versions of the image improperly, leading to a badly blurred image that was useless for any serious study of the object.

Nicks noted the same problems and wrote on Hoagland's web site: "I looked closely at later images such as 80904, and to my distress, I found that the items that I was seeking to observe were greatly changed, or indeed covered over—they quite simply were not the same as on the earlier 80881 image. Of course, I am aware of the difference of elevation of the camera with some of the later images, and the commensurate potential for parallax-type misidentifications. Many such items I have dismissed as natural

Side-by-side comparison of *Viking* lander near-field image with one from *Pathfinder* (1997). Why are the *Viking* era images so much better when *Pathfinder's* camera technology is 20 years newer?

154

or at least not anomalous, because of these very phenomena. Yet repeatedly, items of anomalous interest to me are blurred, changed or covered in later images—images that one would expect to be improving with time, not deteriorating."

The implications of this image tampering and crude manipulation were as abhorrent as they were sickeningly obvious. JPL was clearly trying to hide (and not too well, for some reason...) "something" in these images. It soon became obvious exactly what.

Besides all the near-field *Pathfinder* anomalies, Nicks and Hoagland began studying the so-called super-resolution images of the two distant mountains imaged almost one kilometer away on the horizon of the landing site, the celebrated "Twin Peaks." Nicks, in particular, soon realized that these features showed what he took to be definite signs of engineering as opposed to natural erosional processes. And, although he recognized that the area obviously had been subjected to some kind of catastrophic flood, he could not explain some of the strikingly geometric features he was seeing on the Peaks as simple geology.

For one thing, the "Twin Peaks" were hardly twins. There were obvious, repeating block-like structures on both formations, and some very unusual orthogonal 3D layering on the exposed (downstream) surfaces.

The Twin Peaks seemed to have the definite geometric shapes of pyramids, but highly eroded pyramids at that. According to

Close-up of the right, or northern "Twin Peak." Note layered, geometric, adobe-like structures all the way to the top of the peak.

Nicks, both objects—if they were pyramids—had apparently had their casing literally ripped off in that same massive flood that had devastated the entire area countless millennia before. The debris field in the foreground, which had initially captured Hoagland and Nicks' attention, consisted of a myriad of objects possessing multiple sharp points and edges. They could not then be merely water-eroded (or water-tumbled) rocks. If they were, the sharp edges would have been smoothed out eons before. They either had to be made of much harder materials, or the flood that had happened at this site had been as short-lived as it was catastrophic. Nick's leaned toward the latter conclusion, but it had bigger implications. If the Twin Peaks had actually been not-so-distant arcologies, then the objects that he was seeing in the near-field around the landing site were potentially artificial, metallic *machinery*, ripped from the exposed interiors of the pyramids.

The south "Twin Peak." Note again adobe-like geometric layers of structure on what should be a smooth, water worn hill.

Then, on July 22nd 1997 (Sol 18-19 Mars date), JPL promised to use *Pathfinder's* stereoscopic mast camera (technically the "Imager for Mars Pathfinder" [IMP]) to image the northern twin peak in high resolution. They had used the camera to create the so-called "Presidential Panorama," a (false) color panorama of the area around the landing site. JPL's Timothy J. Parker used some of NASA's specialized enhancement tools to bring out what a JPL press release called the "stratification detail" of the north peak. But when the image was released, it showed only a fraction of the interesting,

NASA image 81977_full.jpg. The official press release for this image reads: "The black and white images show the rightmost of the Twin Peaks in the sharpest view yet. They were processed to bring out the stratification detail in the hill, clearly seen in the image on the right."

layered adobe-like buildings, and a whole lot of sky...

All of which was quite fascinating, and ultimately irresolvable — for without better resolution images from future rover missions, or cleaned-up versions of the *Pathfinder* originals (as opposed to the degraded and obviously sanitized versions deliberately placed on the web by JPL), there wasn't much more that could be done to test the artificial nature of the region. That is, until more than five years later, when NASA introduced a new image enhancement method

The south Twin Peak. Note the layered, block like structures at the top of the Peak, similar to adobe block structures from abandoned Indian settlements in the American southwest.

called super-resolution surface modeling.

In the several years following *Pathfinder*, NASA scientists had been able to take multiple, overlapping images of the *Pathfinder* landing site and enhance them (via a technique called Bayesian interpolation) well beyond their original resolution. This new process revealed that the layered, right-angle patterns on the Peaks bore a strong resemblance to ancient, adobe-hut constructions in the American southwest.

Like the blocky pueblo dwellings of the Anasazi and other ancient peoples, the front faces of the Twin Peaks appear to be stacked one layer upon the other, block upon block, with dark "windows" or entrances clearly visible. The only real architectural differences are the scale—the ones on Mars are much larger—and the fact that the Twin Peaks block structures are built into the face of the pyramids, as opposed to being isolated, stand-alone structures. Either way, the similarity is striking. But, this was not the only surprise to be found on the new super-resolution images.

What had been merely a huge, dark, somewhat blobby shape on the original images suddenly emerged as recognizable on the new, highly processed images. It quickly became apparent that something very interesting might have been imaged on the Martian landscape, between South Peak and *Pathfinder* itself. What it seemed to be—at least to my eyes—was *a sphinx*.

South Twin Peak with large sphinx-like object in the foreground.

158

Like some ancient, strange sculpture, this large object (or, set of objects) almost appeared to be standing guard at the base of South Peak. The clear, geometric, rectilinear relief on the "Peak" was strikingly evident in the new image, confirming Nicks' earlier assessment that it could, indeed, be another shattered arcology on Mars. The "sphinx" lay some distance from the *Pathfinder* lander, at the edge of the previously mapped debris field near the base of South Peak. And, like its counterpart on Earth, it also faced due east—directly toward the Martian equinoctial sunrise. There were even what appeared to be vertically-faced buildings to the left of this potential Martian sphinx. They could easily be viewed as "a temple" (in the Egyptian model), or a distant entrance to the background pyramid arcology itself.

Close-up of the *Pathfinder* "Sphinx" and vertical walls to the left of the main body of the object. See Color Plate #4.

In ancient Egypt, sphinxes were later-Kingdom, much smaller versions of the ancient Great Sphinx on the Giza Plateau. They were routinely used to guard temples, tombs and monuments, much like the Great Sphinx itself guards the three Great Pyramids on the plateau. As a sphinx carries out this task, it is always in the same repose: lying flat on its stomach, forepaws extended outward, ready to pounce into action at a moment's notice. Sphinxes invariably have the head of a man (or woman?) to go with their lion's body. The head, in turn, is framed by the characteristic banded nemyss headdress which is meant to signify the lion's mane.

159

Visual comparison of the Great Sphinx at Giza with the Martian sphinx at the *Pathfinder* landing site.

Even from this compressed narrative perspective, you can see in these NASA-enhanced close-ups that this "Martian sphinx" has all the classic earmarks of its Egyptian counterpart [see Color Page #7] There are two attached and extended forepaws, a body (complete with what looks to be a feline hind leg) and a very clear rounded face, encompassed by a symmetrical nemyss-like Pharaonic headdress. The headdress even has two opposing angles just below the chin, extending outward at about a 45° angle. These characteristics—the extended symmetrical paws and the rounded face framed by that familiar headdress—coupled with its context, lying to the east of a strikingly pyramidal-looking structure, would normally be enough to call this formation's natural genesis into serious question, but when you view it side by side with its counterpart on Earth—the

160

Giza Sphinx itself—the resemblance is uncanny. Beyond that, further enhancement work revealed a twisting "road" that led from the base of the sphinx all the way up to the block-like structures on the South Peak, something that would be needed if indeed the twin Peaks were habitations and not merely very strange looking mountains. And remember—*this* Martian sphinx is guarding these pyramids on Mars at 19.5° N by 33° W.

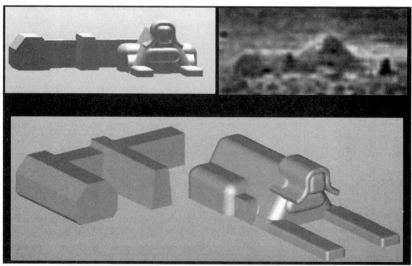

3D CAD reconstruction of the *Pathfinder* sphinx.

The usual suspects—critics who attack everything to do with Cydonia and artificial structures on Mars in general, quickly jumped on the sphinx idea and attacked it. They argued that it was not a distant object, but rather small boulders very close to the *Pathfinder* landing site. They also argued that it did not face east, and that the sphinx and temples were all part of the same structure. In response, I illustrated my interpretation of the structure with a 3D CAD program. This showed how, from the lander's perspective, the "temple" could in fact be a separate structure *behind* the sphinx itself. In any event, my interpretation of the object was not dependent on this. The so-called temple structure could simply be the only remaining walls of an enclosure similar to the one around the Great Sphinx here on Earth. The detractors then switched gears and claimed that the sphinx should be visible in orbital images of the landing site, but that was not true. At that time, images from *Viking* or *Mars Global Surveyor* lacked the spatial definition to resolve an object the size of the Great Sphinx. We would have to wait for *Mars Reconnaissance Orbiter* for that.

Other criticisms included the charge that my arguments were contradictory. Critics point out that it is unlikely that a "sphinx" could survive intact in an area devastated by a large flood. This is simply not a logical conclusion. Ron Nicks, in his geologic analysis of the *Pathfinder* site, concluded that there was indeed a brief but violent (is there any other kind?) flood in the area sometime in the past. He asserted that the front casing wall of the south peak pyramid had been torn off in this process. So how could the "sphinx" have survived relatively intact? Easy. The water flowed around it.

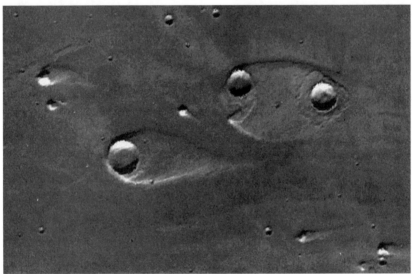

Mars Global Surveyor image of a water flow "teardrop island."

Water, like any other liquid flow, will take the path of least resistance. In images of "teardrop islands" on Mars from various Mars missions, you can see how water flowed around geologic features rather than over them. The same logic applies to the *Pathfinder* landing site. In fact, if the water flow came from behind the Twin Peaks, essentially right at the camera from *Pathfinder's* perspective, as Nicks projects, then objects in the near foreground would have been protected by the presence of the peak itself, which is why it sustained the brunt of the damage by having its front casing ripped off. So it is not contradictory in any way to assume that a possible artificial structure survived intact.

In 2007, the *Mars Reconnaissance Orbiter* ("MRO") finally imaged the *Pathfinder* landing site in high resolution. But, like on so many occasions before, what the image revealed was immediately suspect.

Initially, the image showed promise. The Twin Peaks and the *Pathfinder* lander itself were visible, as was a dark line at the base of the peaks which might have been the sphinx. In aligning the landing site with the peaks, it was clear that the site lines were perfect and matched the panoramas taken from the lander. The dark line was exactly where the sphinx would be if it were really guarding the pyramids.

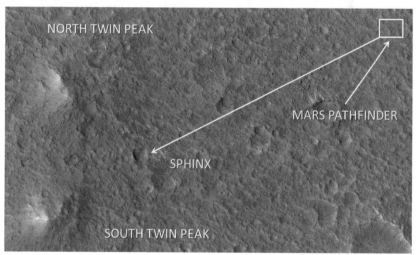

Mars Reconnaissance Orbiter image of the *Pathfinder* landing site showing the Twin Peaks from above. Dark trench in the middle frame may be the "sphinx."

But you looked closer, there were immediate problems. The distinct layered cross-hatching adobe constructions on the fronts of the respective peaks were nowhere to be found, nor was the twisting "road to Damascus" that was so clearly visible in the IMP panorama. In fact, the whole landscape was a soft, monotone gray, with no real detail at all. The peaks looked whitewashed.

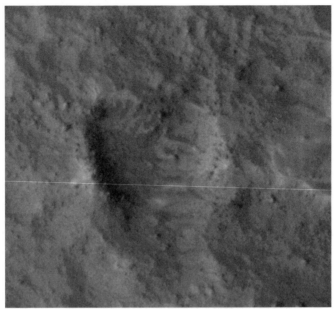

Possible MRO close-up of the *Pathfinder* "sphinx."

It got worse as I zoomed in on the area of the sphinx. Not only were all the structural details missing, they were, unlike how they looked in the panoramas, the same middling, non-descript gray monotone. In this image, the sphinx itself appeared to be nothing but a shallow trench. A quick check of the histograms showed that it had a narrower color band than the overall image. But none of the blocks and structural shapes could be seen. What was worse was that there were not even features between the lander and the "trench" that could have created the optical illusion of a sphinx.

So once again we are faced with only two possibilities. Either the "sphinx" does not exist and neither do the adobe-like habitats on the Twin Peaks, or the MRO image has simply been whitewashed. After careful consideration, I am forced to conclude the latter. Part of this "careful consideration" comes from comparing the NASA supplied super-resolution panoramas with data supplied by the European Space Agency (ESA).

The ESA processing, done mostly by German scientists, showed that the adobe-like construction on the Twin Peaks was definitely valid. Their technique brought out more detail and

Layered, geometric block-like structures on the north Twin Peak from NASA (above) and the European Space Agency (below). Features match.

compared favorably with the work done on the troublesome NASA images. But in the near field, there were numerous discrepancies between specific objects as they had been presented by NASA and as they appeared in the ESA pans.

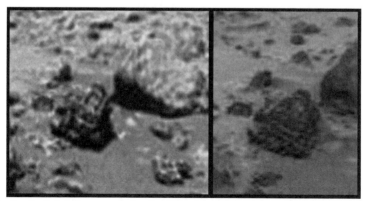

Mechanical debris as seen in the ESA panorama (L) and as presented by NASA (R). Note that some objects are erased completely in the NASA version, and that the mechanical looking object is wallpapered over to look like a natural rock.

Other clearly mechanical objects shared the same fate. A rounded mechanism nicknamed the "Turbo" took on the form of a natural rock when the NASA panoramas were scrutinized.

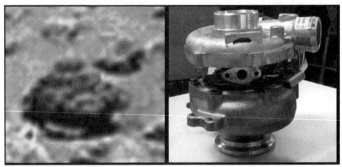

Comparison of the "Turbo" from ESA *Pathfinder* landing site panorama to a terrestrial example.

The "Turbo" from NASA pan image PIA02406.tif. Note how it has changed shape, the obvious misalignment on the Turbo itself and the obvious signs of wallpapering on the image.

At this point, I am forced to go where the data pushes me and conclude that much of the imagery from the *Pathfinder* landing site has been tampered with to cover up the existence of both near field mechanisms and somewhat distant arcologies, notably the Twin Peaks. If in the future ESA or some other agency images the *Pathfinder* site at high resolutions, it is possible that the sphinx controversy, among others, can be settled. But sadly, at the moment, the data coming from the agency simply can't be trusted to make a final determination.

But if the advocates of an Ancient Alien presence on the Red Planet thought that *Pathfinder* had been a roller coaster ride of scientific gamesmanship, it was nothing compared to the political posturing NASA was about to embark on with *Mars Global Surveyor...*

(Endnotes)

1 "CNN Breaking News," live broadcast, July 4, 1997.

2 http://www.enterprisemission.com/planet.htm

CHAPTER 7
MARS GLOBAL SURVEYOR

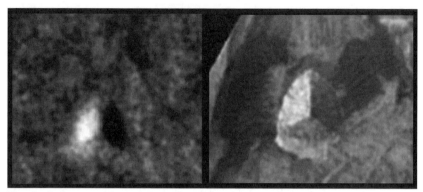

Tetrahedral pyramid on Mars (Cydonia) from *Viking* (L) and *Mars Global Surveyor* (R).

As fascinating as the *Pathfinder* data and all the hijinks around it was, the real prize to the Mars Ancient Alien hunters would be the then upcoming *Mars Global Surveyor*. The successor to the ill-fated *Mars Observer*, it was to be fitted with a slightly upgraded version of that probe's Mars Orbiter Camera (MOC), designed by Dr. Michael Malin at a cost of about $44 million dollars.[1] The camera had a theoretical resolution of between 1.5 and 12-meters-per-pixel under ideal conditions compared to the *Viking* camera's best resolution of about 30

Even though it was based on 1986 technology, Malin Space Science Systems had won the new contract for the camera on *MGS* in what seemed to outsiders to be a less-than-fair competition. Attempts by the military team that ran the *Clementine* mission to the Moon in 1994 to bid on the contract were rejected in spite of the fact that the *Clementine* team's camera was superior in all respects to Malin's. After some complaints to NASA's upper level management, the team asked Dr. John Brandenberg, an old colleague of Richard

Hoagland's from the early independent Mars investigations, to present the camera a second time to JPL's contractually required "open bidding" process. The decision was then abruptly assigned to a selection committee run by JPL, which rejected the *Clementine* camera in a manner that Brandenberg described as "frothing" and instead selected Malin and his camera. This was despite the fact that the Mars Orbiter Camera was basically the same 1986 technology that was on *Mars Observer*. The more advanced and flexible *Clementine* instrument was rejected in spite of the fact that, among other things, it could be gimbaled to point at an "off-nadir" target (a target not directly below the spacecraft), while the entire spacecraft had to be maneuvered to take such an image in the case of the Malin MOC. This meant that precious fuel and orbiter resources had to be used to target specific objects like the Face on Mars. With the *Clementine* instrument, only the camera had to be adjusted. But it also meant that the camera and the images it collected would not be under the control of NASA or JPL, but a separate organization altogether. JPL, it seemed, really wanted their boy Dr. Malin to be the cameraman for the next Mars mission.

With Malin's well-demonstrated hostility toward the Cydonia issue clouding everything around the mission, Dr. Stanley McDaniel's Society for Planetary SETI Research (SPSR) organization arranged a clandestine meeting with NASA's Dr. Carl Pilcher, the acting director of solar system studies, in November of 1997. At the meeting, which was attended by Dr. McDaniel, Dr. Carlotto and Dr. Brandenberg among others, Pilcher feigned interest and promised that Cydonia would be imaged at every opportunity during the "science mapping phase" of the mission. He later dismissed the meeting, saying he "just took the meeting to get SPSR to stop bothering us."

Unfortunately, despite Pilcher's verbal promise that re-photographing Cydonia was now "official NASA policy," SPSR didn't get anything in writing. As the spring, 1998 orbital insertion of *Mars Global Surveyor* approached, Hoagland used the power of his appearances on the *Coast to Coast AM* radio program to ratchet up the public pressure on NASA to formally commit to re-photographing Cydonia. He argued that Malin should not have the god-like power to decide what would be photographed, and that the data stream from the orbiter should be live, as opposed to the up to six months embargo period Malin was allowed under his private contract.

As the public pressure mounted, Malin took to the airwaves himself in an effort to diffuse the situation and retain exclusive control over "his" instrument—which the American public had paid for. Malin chose to give an interview to Linda Moulton Howe, a regular contributor to *Coast to Coast AM*. In the interview, he expressed indignation at the idea that anyone could view him as being responsible for what happened to the *Mars Observer*, claiming that the whole affair had cost him money. He also went to great lengths to claim that getting an image of an object as "small" as the Face on Mars (which is about 2.5 x 2 km) was an iffy proposition at best, comparing it to winning the lottery. When Howe asked him what he would say to those who had waited for almost 20 years for new images of the region, Malin said "...all I can say is, jeez, I'm sorry, that's the reality of the thing."

Of course, this is all baloney. The targeting capability of the *MGS* camera was exceptional, with very little error built into the system. Malin's team had devised an excellent targeting software suite that enabled them to pre-select a Face-sized target with ease. Malin had cited potential targeting errors in the camera's "cross track scanning" as the reason for the uncertainty about getting an image of the Face. Dr. McDaniel had studied this claim in his voluminous 1992 *McDaniel Report*, and declared the following regarding Malin's infamous cross track error claim: "Hitting a specific target the size of the Face is about as difficult as hitting a door with a baseball from a distance of about one foot."[2]

As of this writing, there are no less than seven different *MGS* images of the Face, which I guess makes Dr. Malin a seven-time lottery winner. Obviously, Malin's attempts to deceive the listeners with his "lottery" comments were very worrisome to the Cydonia researchers who hoped to get new and uncensored images of the region. In a page on his website, Malin had made his feelings plain as recently as 1995, when he stated that "no one in the planetary science community (at least to my knowledge) would waste their time doing 'a scientific study' of the nature advocated by those who believe that the 'Face on Mars' [is] artificial."[3]

So the two sides were unfortunately at odds over the idea of simply testing the artificiality hypothesis. As the political pressure mounted, everyone knew that the defining moment for the controversy would be the release of the first Face image. In politics, first impressions are everything, and Malin and JPL had clearly turned the controversy around the Face and Cydonia into a political rather than a scientific debate.

Fortunately, Hoagland orchestrated a fax and email campaign that was aimed at forcing NASA/JPL to target Cydonia early on, during the "science phasing orbit period," when the spacecraft was passing over Cydonia near the Face every nine days. NASA had refused to commit to imaging the Face or any other object of interest in Cydonia until the much later "science mapping phase," and even then, only if they won the proverbial lottery, as Malin had put it. This new campaign finally bore some fruit when NASA announced that they would attempt to target the Face on Mars on April 5th, 1998. Everyone held their breath...

PLAYING IN THE CATBOX

Heading into the April 4th weekend, NASA announced that the Cydonia image targeting the Face on Mars would be released on Monday morning, April 6th at 10:30 a.m. PDT. On the morning of the sixth, everyone in the Mars research community huddled around their computers, anxiously awaiting the image release. Precisely at 10:30 a.m., in accordance with the statement that had been released, the link appeared. But the image that was released was a black, grainy, essentially blank image. Was this a joke?

It was a raw image, supposedly of the data that had been uploaded to the spacecraft. However, even though the image was taken in the full light of midday, it was virtually black. This was not what anyone, even the media, had expected. A few hours later a "processed version" of the image was released by JPL's

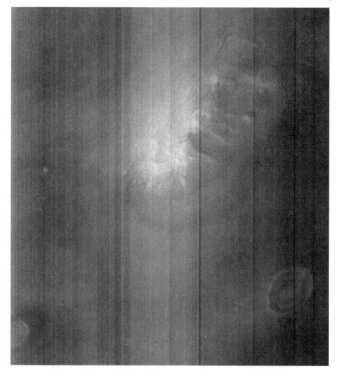

"Raw" Mars Global Surveyor image SP-22003.

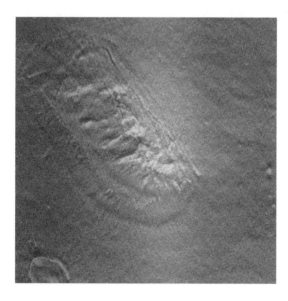

The "Catbox" enhancement of the Face on Mars, reversed (JPL).

Mission Image Processing Laboratory (MIPL). There was only one problem: not only didn't the Face look much like a face, it didn't look like much of anything.

Instead of a high-resolution overhead view of the Face, we got a low-contrast, noisy and washed-out image that was apparently taken well after *MGS* had flown past the Face mesa. This resulted in a view that was looking up from below and to the left side. Details of the right side, previously shadowed in the *Viking* images, were visually compressed by perspective and buried behind the "nose." The image was so lacking in contrast and detail that it gave the impression of a flat, blank desert landscape, with virtually no elevation at all.

This was not what the researchers or the public had expected to see. But it was clearly what JPL and Malin *wanted* the media to show the world. Within an hour of the release of an image that they surely knew was far below the quality of what could be obtained from the raw data, JPL spin doctors had spread out to the various news media pronouncing the Face to be natural. Obviously working from pre-arranged talking points, these spin doctors—employees of JPL mostly—insisted that even though they were NASA scientists they were not speaking for NASA or JPL, but only for themselves. NASA and JPL pronounced that neither would take an official position on the image, thereby draping both organizations in a fallacious robe of objectivity. Surely though, they knew what their employees were doing on their lunch hours, since it was all over the television.

The end result of this was to insulate NASA and JPL from direct criticism on the matter. Any of their employees subsequently found to have made false statements or unscientific arguments over this issue could be dismissed as loose cannons that had acted outside the purview of their responsibilities at the agency. This Clintonian tactic meant that there could never be a second "McDaniel Report" proving NASA's complicity in a campaign of misinformation and ridicule of a scientifically testable hypothesis. At the same time NASA could then claim that it had acted openly and honestly by releasing data quickly and allowing its scientists to comment on the matter.

Members of the independent research community, in their naiveté, were caught off guard by this well-coordinated media assault against them. Still trying to process the raw image themselves and hopefully get a better version of it than had been provided by JPL, they were ill equipped to deal with the media circus around the image release. Facing deadlines, the major media couldn't wait around for Dr. Carlotto or anyone else to process a better version. When the major networks' six o'clock news hours rolled around, they went with the MIPL image.

They were almost uniformly hostile. None other than brilliant planetary scientist Dan Rather pronounced it "a pile of rocks."

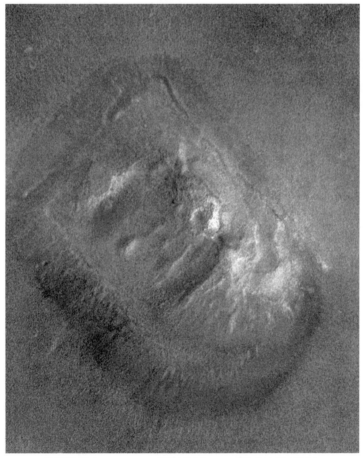

"TJP"eEnhancement of NASA image SP-22003 (JPL).

NBC's Tom Brokaw called the image "proof of what we already knew." Only CNN's John Holliman, who had been friendly to the independent investigation over the years, was somewhat sympathetic, saying that the independent researchers needed more time to properly evaluate the image. He concluded his report by saying "NASA has always said the Face is merely a trick of light and shadow. Some trick." JPL's spin team had done the job.

Then, within *three minutes* of the last national six o'clock newscast sign off, a second image suddenly appeared—again without comment—on the various NASA, MSSS and JPL web and mirror sites. The "TJP" (Timothy J. Parker) enhancement was a significant improvement over the earlier MIPL image. It contained far more contrast and detail, and less noise than the image that had dominated the newscasts.

Parker, a JPL geologist, had produced this second, vastly superior version of the raw data using mostly standard Photoshop tools, and posted his steps on the web. His version had detail that was far more visible and confirmed many Face-like features—including clearly unmistakable nostrils, of all things—but it came too late. Only after the major news organizations had broadcast their reports and made their pronouncements did this considerably improved, much more obviously Face-like image miraculously come to light. Even so, it was still improperly ortho-rectified and gave a less-than-ideal perspective on the object.

That night on *Coast to Coast AM*, host Art Bell was indignant at the political aspects of the day. He considered it a joke that the horrible MIPL version had been the only one available as the TV news had gone to air, and asked Hoagland why it might take another seven hours after producing the MIPL image for Malin's team to release their TJP version. Hoagland admitted that the TJP version should have only taken about thirty minutes to produce, and lamented the fact that the MIPL version made it appear that there was no Face at all. "Well, looking at that image, Richard, I'd have to conclude as well that there is no Face on Mars," Bell said, "and my question now is, where the *hell* did it go?"

Bell summed up the MIPL image by saying that it reminded

him of a pattern his kitty might scratch up in her litter box. It was from that moment forward that the MIPL image would forever be known as the "Catbox" version of the Face on Mars.

Within a day of the release of the new *MGS* image of the Face, there were rising suspicions about its quality. Hoagland found that by comparing it to previously released MOC images, it contained only about 50% of the data that it should have. This deliberate size reduction, along with the fact that SP-22003 contained only 42 shades of gray out of a possible 256, accounted for the egregious amount of noise in the image. It also contributed to the lack of detail and contrast.

Later, Lan Fleming of SPSR tried to recreate the Catbox image with standard software processing tools, to no avail. No matter how hard he tried, he was unable to reproduce the flat, featureless look of the original "enhancement."

Then he decided to try a new combination of techniques. By first applying a high-pass filter (which removes high frequency data from an image) and then a low-pass filter (which removes low frequency data), he got very close to the Catbox look. He then applied a noise filter, which introduced more noise into the image, to reproduce the "graininess" that was so prominent in the Catbox enhancement. But it wasn't until he used an artistic filter called an emboss filter that he replicated the look exactly.

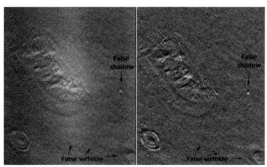

The Catbox enhancement (L) and Fleming's recreation (R).

An emboss filter is a software tool that works by turning lines and edges into a false 3D relief. These edges then become illusory ridges or depressions, depending on the direction that is chosen for the false lighting. This has the effect of creating false visual cues for elevation, effectively scrambling an image to make it less visually coherent. By using these additional filters, the JPL Catbox image was revealed as a simple fraud. As Fleming put it:

"After JPL removed most of the tonal variation in the original image that gives the observer the visual cues to the real three-dimensional shape of the object, they added false visual cues to give the object its rough, jumbled appearance, inadvertently falsifying the appearance of the surrounding terrain as well. The Catbox is not a 'poor' enhancement, as it is often called; it is a crude but very effective fraud perpetrated by employees or contractors to the United States government. Even if the Face is proven to be completely natural, this is inexcusable misconduct and a gross abuse of power. If the Face ultimately is proven to be artificial, the Catbox will certainly come to be regarded as the greatest, most malicious and most destructive scientific hoax since the Piltdown Man, and perhaps of all time."

In other words, in order to get from the original raw MOC SP-22003 image to the eventual Catbox enhancement, which defined the Face to the majority of the public and academia for several years afterward, NASA/JPL/MSSS had gone to the following trouble:

1. Reduced the resolution of the original 2048 x 19200 image strip by 50% to 1024 x 9600, sometime after acquisition of the image;

2. Removed almost 85% of the tonal variations by using high-pass and low-pass filters on the "raw" data;

3. After initial processing, applied another high-pass filter to remove more tonal variations;

4. Applied a noise filter to induce more noise into the image than had already been created by the previous processes;

5. Used an emboss filter to delete visual elevation cues and induce false visual cues into the image.

And all of this, just to discredit an investigation that "no one" at NASA or JPL supposedly took seriously. Just what was it on that original raw data that was so threatening that it would require this degree of suppression? We may never know. But the game was certainly not over.

Although hardly an ideal rendering, the new Face image at least confirmed many of the assumptions and predictions of the early independent investigations. There was indeed a "brow ridge," apparently on both sides and roughly symmetrical. The beveled platform upon which the Face rested could also be confirmed as being close to 98% symmetrical, a condition that was almost unheard of in any natural formation. Beyond that, there seemed to be a curled lower lip, and fairly unmistakable "nostrils" in the nose, right where they should be if they were indeed intended to represent nostrils. There was also a hint of the pupil in the right eye socket.

To Dr. Tom Van Flandern, these obvious secondary facial characteristics were compelling. He argued that such features were inherently predicted by the artificiality hypothesis, and that their existence represented strong enough evidence to conclude that the Face was artificial.

"The artificiality hypothesis predicts that an image intended to portray a humanoid face should have more than the primary facial features (eyes, nose, mouth) seen in the *Viking* images," he wrote on his website. "At higher resolution, we ought to see secondary facial features such as eyebrows, pupils, nostrils and lips, for which the resolution of the original *Viking* images was insufficient. The presence of such features in the *MGS* images would be significant new indicators of artificiality. Their existence by chance is highly improbable. And the prediction of their existence by the artificiality hypothesis is completely *a priori*.

"By contrast, the natural-origin hypothesis predicts that the 'Face' will look more fractal (e.g., more natural) at higher resolution. Any feature that resembled secondary facial features could do so only by chance, and would be expected to have poor correspondence with the expected size, shape, location and orientation of real secondary facial features. Any such chance feature might also be expected to

179

be part of a background containing many similar chance features."

He finished by saying, "In my considered opinion, there is no longer room for reasonable doubt of the artificial origin of the face mesa, and I've never concluded 'no room for reasonable doubt' about anything in my thirty-five-year scientific career."

OTHER MOC ANOMALIES

Fortunately for the Ancient Alien theorists, after the Catbox debacle there were plenty of other images from Cydonia and other regions of Mars to pour through. They revealed that not only was Mars far different than it had been presented by NASA, it was also full of evidence of an Ancient Alien presence.

Some of Dr. Crater's tetrahedrally arranged "mounds" turned out to be tetrahedral pyramids the size of the Great Pyramid itself (see image at opening of chapter). Others were shown to be partially buried dome structures complete with archways and entrances.

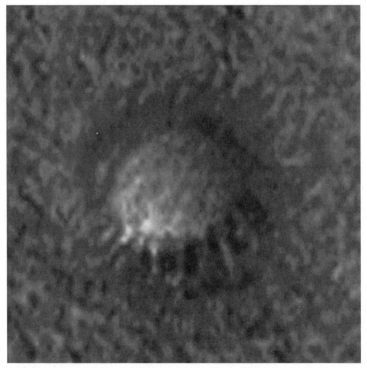

Partially buried mound in Cydonia, near the Face. Note the dome shaped roof, regularly spaced archways and rectangular "door" at the four o'clock position.

Other features elsewhere on Mars looked like helicopter landing pads or multi-layered structures. What separates anomalous objects like this from the natural background are a distinct geometric form, a multi-layered structure (the upper rectangular extension can be seen casting a shadow on the lower, triangular portion), and the presence of a perfectly centered black circle on the upper deck. This dark circle is either painted on the flat surface or it could be a hole in the upper level structure. There is a hint of a cylindrical shape below it. To the left is what almost appears to be a staircase leading to the upper platform. All of these features make it unique against the expected random natural background.

One particularly interesting image was taken in the Ares Vallis region and was described by NASA/Malin as a "grooved or scoured surface in Ares Vallis main channel." In fact, in close-up, it looked like a multi-layered, exposed set of parallel tubes running

The "helicopter pad."

the entire length of the valley. It appears to be an exposed, sub-surface plumbing arrangement for the entire area. If it had once been inhabited, this is exactly the kind of ruined plumbing one would expect to see. NASA has tried to claim these parallel tubes are sand dunes, but some of them are so highly reflective that they appear to be metallic, and in any event it can be plainly seen that they emerge from under the ground on the layer above, and curve back *under* the surface on the lower level, which obviously, "sand dunes" don't do.

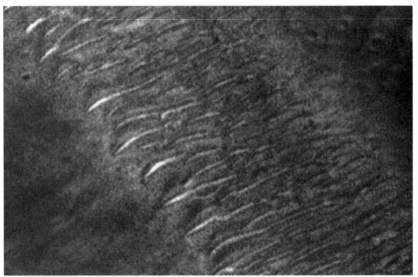

Exposed, parallel metallic tubes from NASA image FHA-00818.

Other images showed even more likely artificial constructs. NASA/JPL image M04-00649, taken of the Coprates Catena region,

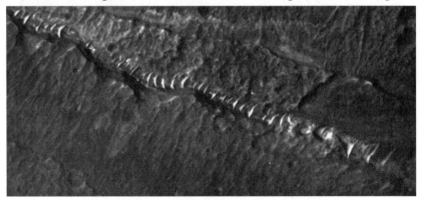

NASA/JPL image M04-00649 in Coprates Catena.

revealed a series of long, tube-like structures supported by regularly spaced, bright (metallic?) supports along the entire exposed length. Again, NASA claims these are sand dunes, but some of them are nearly straight across the gap and they do curve around the shape of underlying tube structures.

Comparison with terrestrial water pipelines is even more illuminating...

Pipelines on two planets. Mars (above) and Earth (below).

One of my favorites among the *MGS*/MOC anomaly collection is a composite image (PIA03914) which NASA describes as a "grooved crater." This "grooved crater" is virtually identical to exposed mineral mines on Earth, as a side-by-side comparison can attest. While the copper mine on the left is undeniably artificial, NASA's explanation for how the "grooved crater" formed is almost incomprehensibly complex:

"The circular feature was once an impact crater. The crater was 2.6 km (1.6 mi) across, about 2.6 times larger than the famous Meteor Crater in northern Arizona. Terra Meridiani, like northern Arizona, is a region of vast exposures of layered sedimentary rock. Like the crater in Arizona, this one was formed by a meteor that

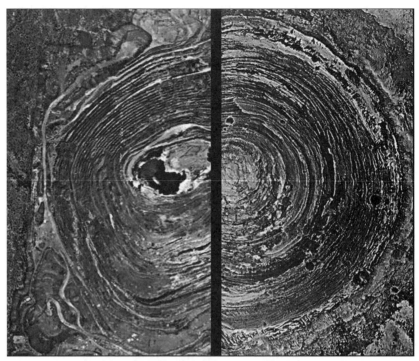

NASA/JPL image PIA03914.jpg and a copper mine in Arizona.

impacted a layered rock substrate. Later, this crater was filled and completely buried under more than 100 m (more than 327 feet) of additional layered sediment. The sediment hardened to become rock. Later still, the rock was eroded away—by processes unknown (perhaps wind)—to re-expose the buried crater. The crater today remains mostly filled with sediment, its present rim standing only about 40 m (130 ft) above its surroundings."[4]

The only problem with this natural model, beyond its twisting complexity and unsupportable assumptions, is that it is simply and obviously wrong. The comparison with Meteor Crater in Arizona is compelling, since that impact was made by a much smaller object. Even so, there is no evidence in Arizona of the "layered sedimentary rock" the NASA caption talks about in reference to the Mars feature. This is because the estimated 10-megaton energy of the Meteor Crater impact would have easily melted any such underlying "layered sedimentary rock." Even if the assumed Mars impactor was smaller or going slower (remember, force equals mass times acceleration) it would, based on the fact that the "crater" is 2.6 times bigger, have generated far more energy than the meteor crater impact, perhaps 30 megatons or more. This would have again melted any underlying, "layered sedimentary rock" that would have been responsible for the grooves. But hey, maybe rock is just harder on Mars than it is on Earth...

In any event, the presence of pipelines, helicopter pads and mineral mines on Mars still paled in comparison with the sexiness of a carved, human Face on Mars and a collection of pyramids. A direct, more overhead view of the Face was still the Holy Grail of the artificiality debate. It took three years after the "Catbox" debacle to get one.

AN EYE FOR AN EYE

On the last day of January 2001, with no warning to anyone in the planetary science community or the independent investigators, Malin Space Science Systems principal investigator Michael Malin released a close-up image of the western half of the Face on Mars. Initially it was very difficult to determine when the image was actually taken, since the normal ancillary data was not linked to the page. It was not until several months later that the ancillary data was actually posted to the page, and it revealed that Malin had taken the Face image back in early March 2000, but somehow he had neglected to include this image in an April 2000 Cydonia data dump.

By not issuing a notice that the image would be taken and then withholding it for almost a year, Malin was once again in violation of NASA's stated policy on Cydonia. In fact, it could be argued that he was in violation on seven counts, since he released six other Cydonia images taken between March 2000 and January 2001 at the same time.

Despite the fact that this was undeniably the best (though partial) view of the Face yet, there were problems. The stated resolution of the image (1.7 meters per pixel) was not exactly the whole story. As with the previous *MGS* view of the Face, there was a large amount of noise in the image, suggesting that the full range of contrast was not made available to the Mars Orbiter Camera. Since actual image resolution is a function of both spatial resolution and contrast range, the actual image resolution is more like five to six

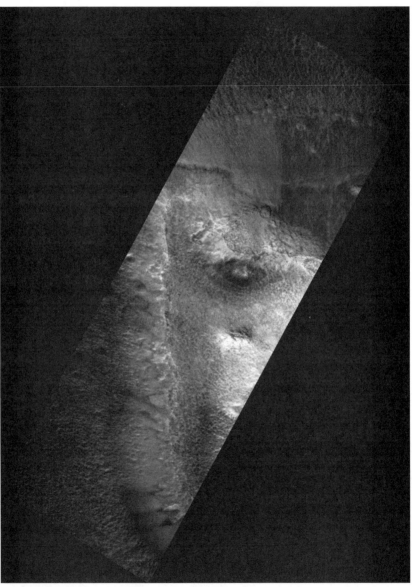

NASA/MOC image M1600184 showing the left, or western side of the Face on Mars. The presence of a brow ridge, a human shaped eye socket and a "pupil" is clearly visible.

186

meters per pixel. This same problem on the previous 1998 Catbox Face image had the effect of reducing the actual resolution to around 14 meters per pixel, as opposed to the stated resolution of around five meters per pixel. What all this induced noise does is make it more difficult to discern the fine structure of a given feature. And the directly overhead sun angle also has the effect of washing out details. That said, this new image was still remarkably revealing.

More than nine years before, former NASA imaging specialist Vince DiPietro proposed that his new analysis of the *Viking* Face images showed the presence of not only what appeared to be an eye socket, but also evidence of a "pupil" of the right size and shape to be a representation of such human features (see Chapter 3). Despite the fact that other researchers using different imaging techniques found the same feature, his prediction was ridiculed at that time by individuals both inside and outside of NASA, and his "bit-slice" imaging technique was roundly criticized. Now, it seemed, DiPietro had the last laugh. The enhancements of the earlier Catbox image had also shown these features, but the low-oblique angle made it difficult to discern if they were truly in the right locations to be evidence of intention on the part of a monumental designer.

But the new image, MOC image M1600184, was a different

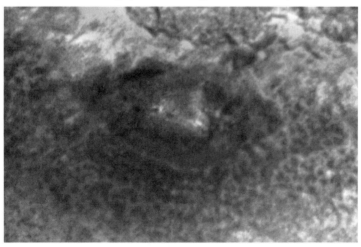

Ultra close-up of the left side "eye" of the Face on Mars showing pupil structure, eyeball and human shaped eye socket. Note regular, repeating pattern of substructures around the eye socket, a sure sign of artificiality.

story. The most noticeable thing about the narrow swath cutting across the center forehead region of the Face, and down across the right eye socket, is that what appeared to be an actual eye socket and pupil in the earlier *Viking* and *MGS* images are, in fact, just that. The "eye socket" was perfectly shaped and positioned to represent a human eye (even including a tear duct) and even though the outlines of the socket are somewhat faded from the sun angle and lack of contrast, it took very little imagination (or enhancement) to determine just what the original shape truly was.

Normally, these kinds of interpretations are dismissed as just that, an interpretation, but this image was further confirmation of Dr. Van Flandern's arguments about secondary facial characteristics. In addition, this new view revealed some incredible architectural details in the fine structure of the Face as well. Around the eye socket was a set of very regular, geometric shapes that appeared to be a sort of honeycomb cellular support structure on the Face itself. This very anomalous and probably artificial pattern is exactly what Hoagland had predicted would be found on the Face when we eventually got a good enough look. He had argued that the Face was not just a Mt. Rushmore-type recarving of an ancient Martian mesa, but a 3D, architectural, "high-tech" construct that with high enough resolution would reveal precisely those kinds of architectural/structural details.

The presence of the "pupil," so controversial previously, could now be placed alongside most of the other *a priori* predictions of the various independent Cydonia researchers. It was now proven beyond a reasonable doubt.

So once again, the higher resolution images had confirmed earlier predictions of the artificiality model. But that was apparently of no interest to NASA. They were too busy getting ready for the next round in what they now viewed as a political war—the release of a direct, overhead view of the full Face itself.

(Endnotes)

1 "An overview of the 1985–2006 Mars Orbiter Camera science investigation." *Mars — The International Journal of Mars Science and Exploration* (Mars Informatics Inc.) 5: 1–60. 6 January 2010. doi:10.1555/mars.2010.0001

2 *The McDaniel Report—On the Failure of Executive, Congressional, and Scientific Responsibility in Investigating Possible Evidence of Artificial Structures on the Surface of Mars and in Setting Mission Priorities for NASA's Mars Exploration Program,* North Atlantic Books (June 1993), ISBN-13: 978-1556430886.

3 http://www.msss.com/education/facepage/face_discussion.html

4 http://photojournal.jpl.nasa.gov/catalog/pia03914

CHAPTER 8
FACE IT, IT'S A FACE

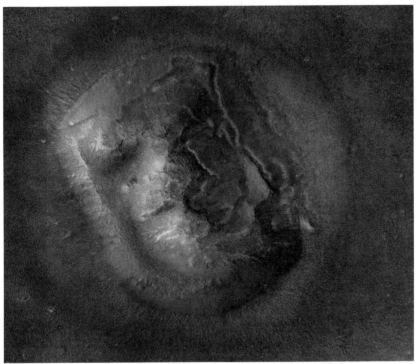

MGS image E03-00824. Right side is distorted and narrower because of improper ortho-rectification by NASA.

In the late morning of the 24th of May, 2001, NASA abruptly released the first *MGS* high resolution and (mostly) overhead image of the Face on Mars. While it was still substantially "off-nadir," taken at an angle off the vertical of 24.8°, as opposed to 45° for the Catbox image, it was a significantly better representation of what the Face would look like from directly overhead. Very quickly, it was also obvious that there were a number of issues with this Face image, as there had been with the Catbox version. The image was the full resolution of 2048 pixels wide, but it was only 6528 long, implying it had been cropped by about two-thirds along the down-track. While it had 175 different tonal variations (compared to only 42 for the Catbox) this still left about 30% of the grayscale information missing. A two meter per pixel spatial resolution was declared for the image by MSSS, which meant that an object as small as a jetliner could be discerned from the data available. Further, it was improperly ortho-rectified, because features that were seen to be along the centerline in *Viking* data and the Catbox image were now skewed to the western side. This had the effect of enhancing the asymmetry of the two sides of the object by stretching the eastern half in proportion to the western side. Overall, however, it was a dramatic improvement over the Catbox image. What was clear from the new image was that while the Face had a substantial general symmetry, it was not a clearly symmetrical human face. What it seemed to be was a half human half *feline* hybrid.

As far back as his U.N. speech in 1992, Richard C. Hoagland had asserted that those who expected the Face to be fully human were wrong in their assessments. Using then primitive computer technologies, he had done a series of symmetry studies of the *Viking* data and concluded that this human/feline hybrid appearance might have been intentional.

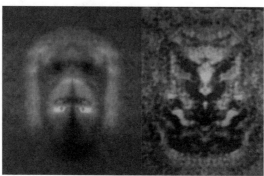

Left and right side Face symmetries from *Viking* data (courtesy Richard C. Hoagland).

The new image seemed to confirm this model, as updated symmetries were done and the human/lion impression persisted.

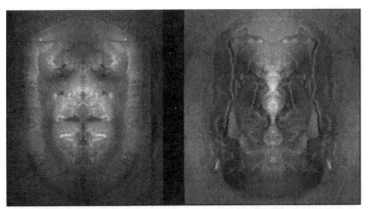

Left and right side symmetry studies from NASA image E03-00824.

Despite another prediction of the artificiality model being fulfilled, NASA was determined to keep this new Face image from gaining traction with the public or news media. They released the new image amid a flurry of extremely negative public comments simultaneously posted on several official NASA websites. Specially prepared political hit pieces on the Face were posted coincident with the release of the new image. Titled "Unmasking the Face on Mars" and authored by NASA without a byline, the article series resorted to gross distortions and outright fallacies in its political attack on E03-00824.

In "Unmasking the Face on Mars," NASA used all the standard debunking and propaganda techniques they had honed over the previous 20 years of debate on the Cydonia issue. They described the Face as a "pop icon," never mentioned the existence of any of the other anomalies in the Cydonia complex like the D&M or the Tholus, pretended the geometric relationship model didn't exist and used a cartoon to ridicule the idea that the Face was anything other than a common Martian mesa. Jim Garvin, chief scientist for NASA's Mars Exploration Program, was quoted as saying that the Face reminded him of Middle Butte Mesa in Idaho. Of course, the article didn't contain an image of Middle Butte, making it impossible for anyone to assess NASA's integrity when making such comparisons. Several industrious independent researchers were able to track down a satellite image of Middle Butte however,

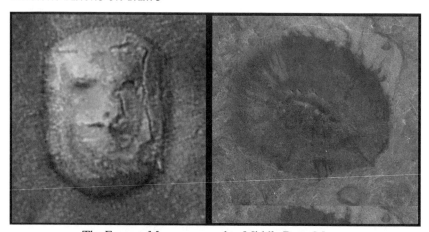

The Face on Mars compared to Middle Butte Mesa.

and they were quickly able to show that Garvin's comparison was absurd.

Any reasonably observant person could easily conclude that Middle Butte bore no resemblance to the Face. For one thing, the Face had two equal length parallel straight edges on either side of the base that ran straight for hundreds of meters on either side. Middle Butte is just a common cinder cone, with no parallel edges. It also did not have two eye sockets, two brow ridges, a nose with nostrils or a "mouth" that went through. But NASA had apparently decided that simply making silly comparisons wasn't enough. Later in the piece, they showed a vertically compressed, grossly distorted and upside down version of the Face, supposedly generated by a shape-from-shading algorithm. It was so badly distorted that parallel features clearly visible in the *Viking* overhead shots from 1976 ended up completely divergent. NASA used this ginned-up version of the image to try and confuse readers, and in some cases, it worked. The reality is that it is highly unlikely that anyone would recognize a picture of their own grandmother if it was stretched horizontally, flattened, compressed and shown upside down.

"Unmasking the Face on Mars" then went on to link this deliberately distorted image to a very impressive-looking 3D color version of separate data from the *MGS*/MOLA (Mars Orbiter Laser Altimeter) instrument—and then claimed that these two images taken together "proved" that the Face on Mars was just another

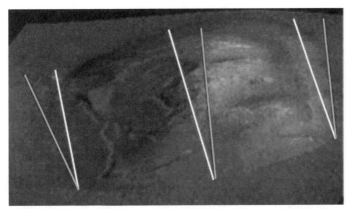

Supposed MOLA generated image showing how NASA stretched, flipped and distorted the Face on Mars to make it appear random and natural. Image is actually derived from MOC image E03-00824.

Martian hill. According to the story, "The laser altimetry data are perhaps even more convincing than overhead photos that the Face is natural. 3D elevation maps reveal the formation from any angle, unaltered by lights and shadow. There are no eyes, no nose, and no mouth!"

There was one more major problem with NASA's argument: the MOLA instrument they were relying so much on had a resolution of *150 meters per pixel*. So NASA was basing their entire "it's just a hill" argument on a MOLA image that was *six times worse* than the original 25-year-old *Viking* data. At that resolution, an object has to be about the size of three baseball stadiums to even show up. To argue that there are "no eyes, no nose, and no mouth!" based on such a crude instrument is not only scientifically absurd—it is scientifically dishonest. The simple truth is that the MOLA is incapable of resolving a feature even as small as the pupil DiPietro had found. And the old-fashioned MOC camera, with its 1.5-meter spatial resolution, and 256 shades of gray scale resolution, is thousands of times more accurate.

It was later discovered that the image that NASA was passing off as "MOLA generated" data was not even from the MOLA instrument itself, but was just a deliberately "de-resed" version of the MOC image. It had been created using Photoshop tools by a NASA contractor named Jim Frawley and Garvin himself, and by

Frawley's own admission was a near complete fabrication.[1] Just why NASA would have to go to all this trouble and deceit to discredit an object on Mars that they claimed they have no scientific interest in remains a mystery. The reality is that the MOLA claims were not only false, but they were calculatingly designed to "scotch this thing for good," as one unnamed JPL scientist had put it after the 1998 Catbox fiasco.

Fortunately, this time NASA's lies about the Face were exposed quickly, and by the time the new image had been released the Internet had become a better arbiter of truth than the mainstream news networks. And there were soon many more images of the Face and Cydonia to come, from sources beyond the compromised NASA food chain.

On Friday, April 12, 2002, a new image of Cydonia and the Face on Mars from a probe named *Mars Odyssey* was released by NASA. Run by Arizona State University's Dr. Philip Christensen, the *Mars Odyssey* probe carried the potential to unlock a great many of the mysteries about Cydonia once and for all. While it carried a very good visual image camera, the probe was built around an instrument named THEMIS which was capable of taking infrared images of Mars, something only the Russian *Phobos 2* probe had been able to accomplish previously. It also had a five-band color camera, and offered the hope that we might someday get color images of Cydonia. Unfortunately, we did not get a multi-spectral nighttime infrared image of the Face and surrounding structures, nor did we get a full-color image of the Cydonia complex in the promised five-band color. What we did get was a nice grayscale strip from the spacecraft's visible light camera. At 19 meters-per-pixel, the image was substantially better resolution than the 50 meter per pixel images taken in the *Viking* era. Still, it was an order of magnitude lower than the four to five meters per pixel images obtained in the best of circumstances from *Mars Global Surveyor*.

This did not prevent it from being useful. It was still a better, more direct overhead view of the center of the Cydonia complex than we had ever had before, allowing the independent researchers to view the specific objects of interest in context and at a resolution not generally seen. The D&M, for instance, had been almost completely missed in the publicly released images from Malin Space Science Systems. Previously identified objects of interest that were visible in the new *Odyssey* image strip included the Face, the D&M Pyramid, the Fort... and some surprises.

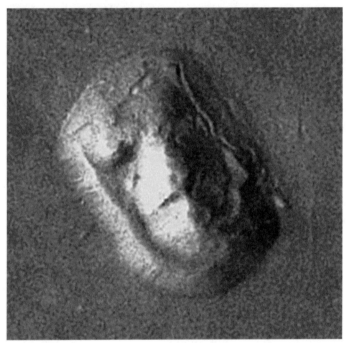

Mars Odyssey image of the Face on Mars.

The first thing noticed was a so-called "massive tetrahedral ruin" adjacent to the symmetrical mesa just south of the Face. This object was first spotted on the infamous Catbox *MGS* image strip. It had been stylistically interpreted from previous images as a "dolphin," or various other absurd pictogram shapes, with one amateur pixologist even claiming he saw a "trailer park" at the base of the object. Part of the illusion was the extreme forced perspective of that original *MGS* image—taken 45° off-nadir—which effectively distorted the shape of the ruined pyramid. We could now see clearly from directly overhead that there were two distinct faceted walls that once made up this ruined tetrahedral structure. The object lies just south of a suspiciously symmetrical "mesa"

197

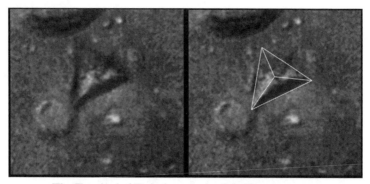

The Tetrahedral Ruin just south of the Face.

and—conveniently—lies at exactly 19.5° off the base symmetry axis of the D&M. This same axis passes right between the eyes of the Face.

Of all the objects captured on the frame, however, the real prize was the D&M. One of the most controversial objects in the entire Cydonia artificiality debate because of its status as the lynchpin of the Hoagland/Torun Geometric Relationship Model, the D&M had always been crucial to deciphering the correctness of the original Cydonia observations from two decades previous. The most crucial question surrounding the D&M had been that of the pyramid's five-sided symmetry, which was suggested strongly by the original *Viking* data and was an issue of some contention in years past. Vince DiPietro, for one, had maintained for years that the D&M was only a four-sided object, objecting to the implication of a "fifth buttress" in the shadowed side of the pyramid on the *Viking* data.

This projected five-sided symmetry held up extremely well in the new image. Especially notable were the four clearly defined faceted sides to the pentagonal pyramid meeting at a central apex— exactly as observed by Hoagland and Torun on the original *Viking* data in 1989. The new image also (despite deep shadow) could verify the existence of a "fifth buttress" to the northeast—the final piece needed to complete the pentagonal form and reconstruct the object's original, undamaged shape. The buttress seemed to be pretty much the same length as the other visible buttresses (the southeastern buttress is mostly buried under debris flow from the mild collapse the object has endured) and verified the predicted geometric form

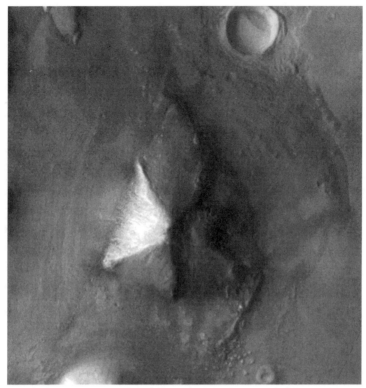

The D&M pyramid from *Mars Odyssey* 2001 (THEMIS-ASU).

proposed by Torun in 1988 perfectly. Obviously, such a "hit" was way beyond even "the power of randomness," and was a compelling confirmation of the validity of Torun's original work.

The other major discovery from the new image was that the entire structure could now be seen to be placed atop a huge rise (or platform) much like the Giza Plateau on Earth—and this newly discovered platform, which was not resolvable in the *Viking* data, seemed to have a geometry all its own.

When you rotate the image of the D&M from the way we are all used to looking at it, we can see that there are two distinct but partially buried edges to the plateau that the D&M rests on. These two edges, not visible in the original *Viking* data, meet in an apex point that is exactly aligned with the SE buttress on the opposite side of the structure.

What this new perspective on the D&M now allows us to see for the first time is that the pyramid rests on a 2D, *seven-sided*

199

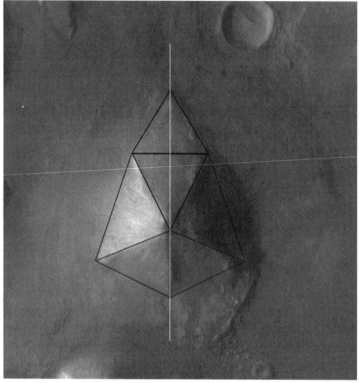

Alternate, secondary symmetry of the D&M pyramid on Mars (THEMIS-ASU).

platform (or base) upon which the massive 3D five-sided "Rosetta Stone" structure was constructed. It also revealed an additional, second alternative line of symmetry for the object, which conversely produced a second, bilaterally symmetrical four-sided geometric figure.

When these two shapes (the seven-sided platform and the five-sided pyramid) were superimposed upon each other, they once again reinforced the quintessentially tetrahedral message of Cydonia. Indeed, one of the new internal angles generated by the new figure is none other than the ubiquitous 19.5°. Not only did this new data flatly validate the original Torun reconstruction and analysis of this enigmatic object, but it also demonstrated the correctness of the geometric relationship model derived from it.

The new image also allowed researchers to do an actual side-by-side-by-side comparison of the Face from *Viking*, *Mars Global Surveyor* and *Mars Odyssey*. Immediately, several things became

200

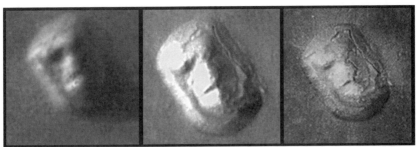

Three views of the Face on Mars: *Viking* (left) *Mars Odyssey* visual image (middle) and *Mars Express* (right). These are the three most directly overhead views of the Face.

obvious. First, the *Odyssey* image confirmed that the *MGS* April 2001 image had been poorly ortho-rectified, as the Face platform was much narrower and more symmetrical than it was on the *MGS* image. Also, the "nostrils" from the Catbox image had returned after being nearly invisible on the April 2001 image and the "lion"-side eye socket appeared to be better aligned with the opposite socket.

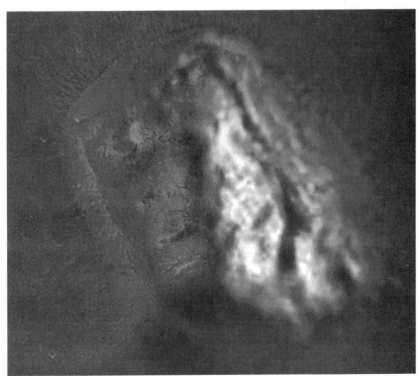

Composite image made from May 2001 *MGS* image and July 2003 *Mars Odyssey* five-band color image. Image courtesy Richard C. Hoagland.

201

All of this only added to the belief among the independent Mars researchers that they'd been right all along about Cydonia. But THEMIS, ASU and *Mars Odyssey* were about to give us an even more incredible confirmation of the model than all the pictures of the D&M and the Face combined. We were about to get an infrared image that would shake up the debate decisively.

The image above is an enhanced close-up created by Richard C. Hoagland from a combination of three 2001 *Mars Odyssey* visual frames taken by the *Odyssey* visual camera and the 2001 *Mars Global Surveyor* image of the Face on Mars (E03-00824). The *Odyssey* image release is officially designated JPL/ASU V0 3814003. The five frames—from the near-infrared end of the visible spectrum to the violet—were acquired by the *Odyssey* spacecraft as it flew over the Cydonia region on October 24, 2002, precisely one year after it arrived in Martian orbit.

What made this color close-up so remarkable (see color page #6) is that for the first time in over a generation, a NASA spacecraft had acquired multi-spectral images of the Face as seen in morning light. This meant that the illumination was coming from the east, whereas in all previous images the light was coming from the west, leaving the right, or "eastern" side, in shadow. What this unique sun angle revealed was nothing less than revolutionary. Even casual examination of the Face as seen in this new light revealed two new pieces of vital information:

1. The eastern side, under even this pre-dawn illumination, is incredibly reflective, and
2. In lowered contrast images, the source of this anomalously high albedo is reflecting off a series of highly geometric panels.

The image was taken by the *Odyssey* camera at 4:39 a.m., local Martian (Cydonia) time, well before dawn. Further reading of the data that was published with the image revealed that the "phase angle"—the geometric relationship between the sun, the Martian surface directly underneath the spacecraft and *Odyssey* itself—was 90.3°. Since 90° would indicate the sun was literally on the eastern horizon, the slightly greater angle reveals that actually the sun was 0.3° below the horizon when the image was acquired and even slightly lower at the location of the Face itself. The last line in the table, "Description: Cydonia—Face at Night," confirms this geometry. Technically, then, this *Odyssey* dawn image was actually acquired just before sunrise, with the sun still hidden below the Cydonia horizon. This simple, inarguable geometry marks the high brightness of the Face's eastern side—before the sun has risen—as extraordinary. This, in turn, leads directly to the pivotal question: just what could make an average Martian mesa so incredibly reflective?

NASA image V0 3814003.

A side-by-side comparison revealed the true incongruity of such a brilliantly glowing object in the pre-dawn light. The official NASA version of the Face from V0 3814003 (left) is totally washed out on the illuminated (eastern) side—even though the image was shot before the sun had risen. In the Hoagland enhancement (right) some surface details can just be seen beneath the glare. Decreasing the brightness by about 90% revealed some astonishing 3D geometry on the Face's eastern side.

What this establishes is that the Face is reflecting on the order of 99.9% of the sunlight hitting it in this image, while the rest of the "normal" Martian surface is only reflecting about 20%. That means that the reflective, eastern side of the Face is not only made up of something different from the surrounding landscape, but that this "something" can only be a highly reflective material—like metal. The overall effect would have been identical to an internal lighting system, producing an imaging effect almost like an x-ray, making the internal architectural structure of the Face on Mars visible for the first time.

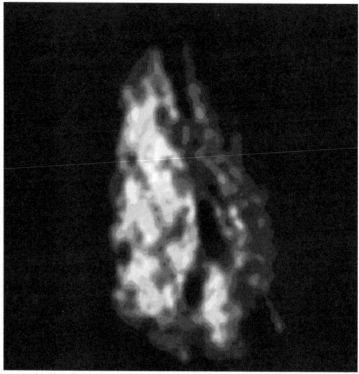

Geometrically aligned panels on the right (eastern) side of the Face on Mars.

A close examination of the image shows that the right side of the Face has an underlying geometric set of panels covered by only a hint of dust. The infrared character of this image allows us to "see" details under this light dust covering. The panels are massive, hundreds of meters in some cases, and we can be sure they are not artifacts of the enhancement process because of the fact that they align with the north/south axis of the Face instead of the vertical/horizontal axis of the image scan.

Put simply, there is no way that a simple mesa—whether it is on Mars or Earth—can have such a self-luminous, underlying geometric structure. If there were any doubt remaining about the Face being artificial, this one image should have ended it. But yet, there was still one last test to take and one more hurdle to jump. What would the ultra-high resolution of the new *Mars Reconnaissance Orbiter* reveal?

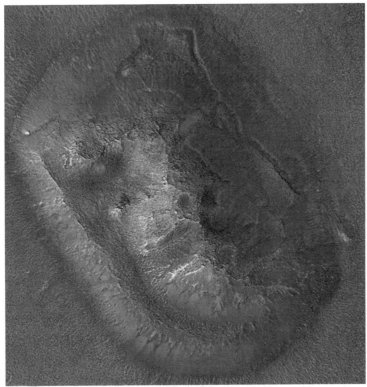

MRO image PSP_03232_2210_RED

Mars Reconnaissance Orbiter

On August 12, 2005 NASA launched the advanced *Mars Reconnaissance Orbiter* toward the Red Planet. It took seven months to reach Mars and another five months of aero-braking maneuvers to achieve a stable science orbit around the planet. *MRO* was equipped with a variety of instruments including the new "HiRISE" camera—the High Resolution Imaging Science Experiment. HiRISE had a 19.5-inch-wide telescopic mirror (yes, there's that number again), and CCD camera capable of acquiring images over 20,000 pixels wide at a resolution of up to 0.3 meters per pixel under ideal conditions. This translates out to an 800 megapixel image capable of resolving objects on the Martian surface as small as the *Pathfinder* rover. The new camera was so good that NASA's Jim Garvin called it a "microscope in orbit."[2] *MRO* also contained instruments to measure temperature, pressure, water vapor and dust levels as well as a powerful ground penetrating radar instrument named SHARAD.

Of course, *MRO* images of Cydonia and the Face were high on the priority list of the independent Mars researchers, and NASA was for once fairly quick to respond. In April 2007, shortly before we went to press on *Dark Mission*, NASA released a high resolution HiRISE image of the Face on Mars—PSP_003234_2210. At slightly less than 11 inches per pixel compared to previous *Mars Global Surveyor* images of some 4 feet per pixel, the *MRO* image represents by far the most detailed image ever taken of the Face.

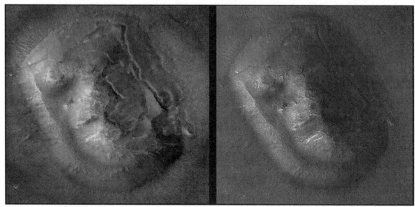

Side-by-side comparison of the April 2001 MGS images of the Face on Mars (left) and the April 2005 *MRO* image (right).

At first glance, the *MRO* image didn't really look much different from the earlier *Mars Global Surveyor* images of the Face. When zoomed out to show the entire Face, the by now familiar features were still present and accounted for. The brow ridges, the eye sockets, the pupil, the nostrils, the tear duct and the parallel base of the object were clearly visible. The *MRO* image was more accurate because it had been taken more directly overhead like the *Mars Odyssey* and *Mars Express* images. As a result, it was less distorted due to improper ortho-rectification and was much closer to those images in terms of accuracy. But as they say, the devil is in the details, and that is where *MRO* and the HiRISE camera excelled. The detailed close-up views of the Face from the new image were astonishing.

If, as the theory goes, the Face on Mars is artificial, then in extreme close-up we would expect to see evidence of structural composition. Things that look like rooms, walls, exposed beams, girders etc., should all be readily apparent. We would also expect to see some confirmation of earlier observations, like the panel-like structures on the right side of the Face seen in the THEMIS pre-dawn infrared image.

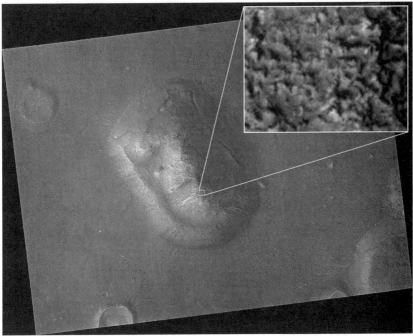

Sectional enlargement of the "nose" area of the Face on Mars from *MRO* image PSP_03232_2210. Note the eroded walls running at right angles to each other.

And that is exactly what we got.

The key to proper interpretation of aerial or satellite imagery of man-made ruins on Earth lies in noting the multiple examples of parallel walls and repetitive, right-angle geometry. As Carl Sagan once put it, "Intelligent life on Earth first manifests itself in the geometric regularity of its constructions..." Natural geologic features do not display these repeating 90-degree relationships, at least not on a large scale or repetitive basis. Right angled enclosed rooms, repeating linear wall alignments and redundant examples of geometrically organized uniform-width features are the exclusive domain of intelligently-designed ruins.

Right-angled wall on the Face on Mars compared to Roman fortifications in Britain.

In this enlargement one can easily discern row upon row of obviously collapsed geometrically ordered ruins. The arrangement includes a stair-stepped connected "wall" descending through the center of the image, as well as a host of other, equally rectilinear ruins around it. The wall bears a strong resemblance to eroded Roman fortifications like "Hadrian's Wall" in Britain as seen from orbit.

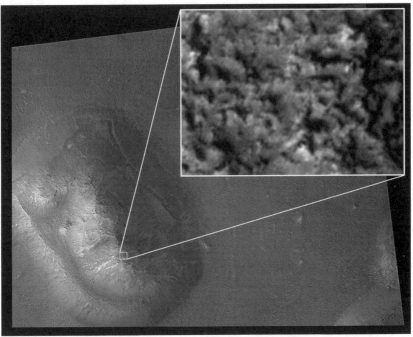

Close-up of geometrically organized ruins from the "chin" area of the Face.

Examining another region of the Face—this one approximately a mile from the first location, reveals another type of equally obvious construction further down at the base of the "chin" area. These square, cellular-like features are strongly reminiscent of sand filled Middle Eastern ruins and upturned buildings seen in shattered earthquake zones on Earth.

The fact that these ruined structures look somewhat different from the previous example is due to two reasons. 1) The original architectural geometry was simply different, composed of larger and deeper individual "cells" consistent with a stouter structural foundation for the vast mass of the entire Face; and 2) the ruins' current physical location on the flat platform area at the base of the chin slope allows all the eroded debris from higher on that slope to cascade down and into the deep geometric cavities of the rooms.

If you look carefully at the top portion of the above image, you will notice a bewildering number of straight lines, sharp edges, 90-degree angles, flat sides and more parallel-width walls. These are all unnatural features, never seen on any ordinary "geological"

209

Collapsed apartment building on Earth compared with square "cell" on the Face on Mars. Scales are identical.

formation, and certainly not in such extraordinary numbers and close, repetitive association.

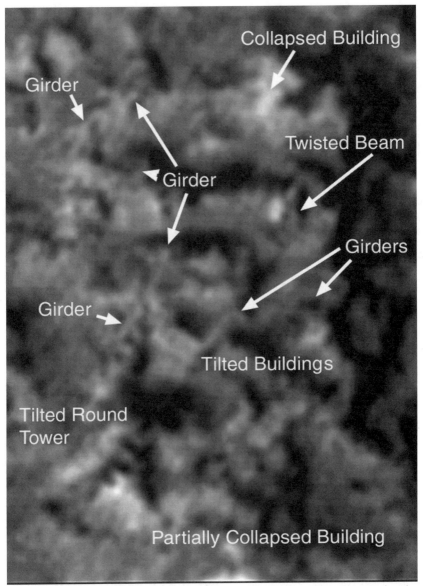

Close-up of structural ruins from the "chin" area of the Face. Annotation by Richard C. Hoagland (used with permission).

The truth is, nothing in this sectional enlargement is natural. Natural geology doesn't come with parallel walls, multiple 3D planes, twisted beams or repetitive examples of obvious "girders." High-tech structures always do. These are nothing less than the inevitable hallmark of closely-associated, shattered and eroding high-tech structures, whose mere presence on the exposed surface of the Face is an overwhelming confirmation of its completely artificial nature. These same evident geometric relationships are found in any modern city, the inevitable consequence of the construction of repetitive, multi-storied structures formed from basic geometric units. You can even see the beams and girders that we would expect in any sufficiently high-resolution image of this ancient, high-tech structure known as the Face on Mars.

And as you examine the *MRO* image even more closely, there is yet more to find. On the shadowed right side of the Face, exactly where the THEMIS infrared image of the Face showed glowing, metallic panels in the pre-dawn light, we find further evidence that the Face is a constructed artifact.

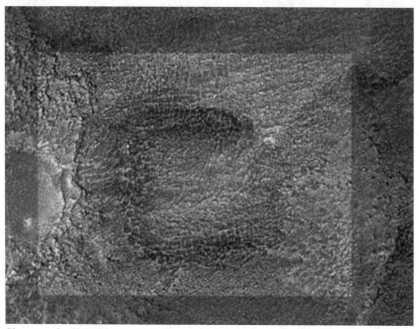

Close up of cellular, honeycomb-like structures on the right side of the Face on Mars.

Just below the "chin" and in the same area of the glowing panels is a collapse feature hundreds of feet across. Regular, repeating cell-like structures can be seen, probably exposed by the collapse of the "roof" into the middle of the feature. This is another undeniable example of non-random, non-fractal erosion. It is exactly what a mesh-like substructure exposed by a structural failure would look like from above.

Collapsed stadium on Earth.

The truth is, any structure, no matter how high tech and how well designed and constructed, will eventually collapse under its own weight if left long enough and allowed to fall into a state of disrepair. Our own civilization, if abandoned for a thousand years or so, would erode quickly and completely disappear from the face of the Earth. What's extraordinary about the Face on Mars is that it has probably been abandoned for far longer than that, perhaps as long as 1.35 million years, and yet the signs of the majesty in which it once reveled are still visible for all to see.

The fact remains that after 30 years of back and forth debate about Cydonia between the independent researchers and the NASA/media establishment, the mainstream argument still comes down to the same thing it did in 1976: "It's not a Face."

Or, more accurately:

"It's not a Face, in spite of the fact it rests on a bilaterally symmetrical platform, it has two aligned eye sockets, the tip of the nose is the tallest point on the structure, there are two clearly defined

213

nostrils in the nose, the west eye socket is shaped like a human eye including a tear duct, there is a spherical pupil in the eye, there are rectangular, cell-like structures around the eye, the two halves of the Face make up two distinct visages when mirrored, one human, one feline, it is placed nearby a series of pyramidal mountains which have rectilinear cells visible in their interiors at high resolution, it is in close proximity to an isolated pentagonal "mountain" which is bilaterally symmetrical about two different axes, it has anomalous reflective properties under pre-dawn conditions, it is surrounded by a series of tetrahedral mounds which are placed according to tetrahedral geometry and in high resolution it displays features identical to ruined artificial structures here on Earth..."

Well, I could literally go on and on. But you get the point. Their argument is really weak.

Whether you look at the Face from far away, like the early *Viking* images, or in ultra-close-up, as in the *MRO* image, you still see one thing clearly—someone built it. It is not an example of the mythical "pareidolia" that the NASA sycophants are always asserting. It is rather an elegant and enduring example of something far more: an Ancient Alien extraterrestrial monumental architecture.

And that is, after all, "some trick..."

(Endnotes)

1   *Dark Mission – The Secret History of NASA*, chapter 8.

2   "NASA Outlines Mars Missions." Space.com. Retrieved July 4, 2006

EPILOGUE

When I first began this project, it was my intention to cover far more examples of Ancient Alien ruins on Mars than just famous objects in Cydonia and the other sites we have discovered here. Unfortunately, the Face and other enigmatic landforms there have a way of taking over a manuscript and dominating it. That said, because of time and space constraints, I was unable to even cover all of the information on that region that I was planning to include. There are so many more examples of the vast Type-II Ancient Alien civilization that once flourished on the Red Planet that I have been forced to retrench and conclude that the only way to do the work of researchers like George Haas, Joseph Skipper and many others justice is to move many of those objects into *Ancient Aliens on Mars II* for next year. While this disappoints me to some extent, it will give me the freedom and time to do justice to mysteries like the petroglyphs on Mars, the Martian skull, and the enduring mysteries of the rovers Spirit, Opportunity and Curiosity. There are simply so many more Ancient Alien stories to tell about Mars that I can't wait to get started, even as I close the proverbial chapter on this work.

Between now and then, I hope you will tell me what you think of what you've read so far either through my social media pages or in your Amazon.com reviews.

One last thing I wish to address. Much has been made by critics and debunkers of the quality of the images in both this book and *Ancient Aliens on the Moon*. The simple fact is that print simply cannot properly display the images presented herein. So by the time you read these words, all of the images presented here and in that previous book will appear on my Picasa web album in full color and full resolution. The web link is:

https://picasaweb.google.com/108348453161239832103.

The bottom line is that Mars is a place of such rich mysteries that this series really could go on for a long time. I can't wait for that journey to continue and I hope you'll be joining me at each turn.

Mike Bara— 9/18/2013

Epilogue

Mike Bara is a *New York Times* Bestselling author and lecturer. He began his writing career after spending more than 25 years as an engineering consultant for major aerospace companies, where he was a card-carrying member of the Military/Industrial complex. A self-described "Born Again conspiracy theorist," Mike's first book *Dark Mission* (co-authored with the venerable Richard C. Hoagland) was a New York Times bestseller in 2007. *The Choice* from New Page Books was published in 2010.

His 2012 book, *Ancient Aliens on the Moon* was a cult best-seller and created a firestorm of controversy.

Mike has made numerous public appearances lecturing on the subjects of space science, NASA, physics and the link between science and spirit, and has been a featured guest on radio programs like Coast to Coast AM with George Noory. He is a regular contributor to the History Channel programs *Ancient Aliens* and *America's Book of Secrets*, both of which are now showing on the H2 channel.

ANCIENT ALIENS ON THE MOON
By Mike Bara
What did NASA find in their explorations of the solar system that they may have kept from the general public? How ancient really are these ruins on the Moon? Using official NASA and Russian photos of the Moon, Bara looks at vast cityscapes and domes in the Sinus Medii region as well as glass domes in the Crisium region. Bara also takes a detailed look at the mission of Apollo 17 and the case that this was a salvage mission, primarily concerned with investigating an opening into a massive hexagonal ruin near the landing site. Chapters include: The History of Lunar Anomalies; The Early 20th Century; Sinus Medii; To the Moon Alice!; Mare Crisium; Yes, Virginia, We Really Went to the Moon; Apollo 17; more. Tons of photos of the Moon examined for possible structures and other anomalies.
240 Pages. 6x9 Paperback. Illustrated.. $19.95. Code: AAOM

ANCIENT TECHNOLOGY IN PERU & BOLIVIA
By David Hatcher Childress
Childress speculates on the existence of a sunken city in Lake Titicaca and reveals new evidence that the Sumerians may have arrived in South America 4,000 years ago. He demonstrates that the use of "keystone cuts" with metal clamps poured into them to secure megalithic construction was an advanced technology used all over the world, from the Andes to Egypt, Greece and Southeast Asia. He maintains that only power tools could have made the intricate articulation and drill holes found in extremely hard granite and basalt blocks in Bolivia and Peru, and that the megalith builders had to have had advanced methods for moving and stacking gigantic blocks of stone, some weighing over 100 tons.
340 Pages. 6x9 Paperback. Illustrated.. $19.95 Code: ATP

THE ILLUSTRATED DOOM SURVIVAL GUIDE
Don't Panic!
By Matt "DoomGuy" Victor
With over 500 very detailed and easy-to-understand illustrations, this book literally shows you how to do things like build a fire with whatever is at hand, perform field surgeries, identify and test foodstuffs, and form twine, snares and fishhooks. In any doomsday scenario, being able to provide things of real value—such as clothing, tools, medical supplies, labor, food and water—will be of the utmost importance. This book gives you the particulars to help you survive in any environment with little to no equipment, and make it through the first critical junctures after a disaster. Beyond any disaster you will have the knowledge to rebuild shelter, farm from seed to seed, raise animals, treat medical problems, predict the weather and protect your loved ones.
356 Pages. 6x9 Paperback. Illustrated. $20.00. Code: IDSG

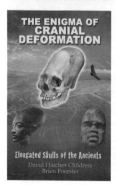

THE ENIGMA OF CRANIAL DEFORMATION
Elongated Skulls of the Ancients
By David Hatcher Childress and Brien Foerster
In a book filled with over a hundred astonishing photos and a color photo section, Childress and Foerster take us to Peru, Bolivia, Egypt, Malta, China, Mexico and other places in search of strange elongated skulls and other cranial deformation. The puzzle of why diverse ancient people—even on remote Pacific Islands—would use head-binding to create elongated heads is mystifying. Where did they even get this idea? Did some people naturally look this way—with long narrow heads? Were they some alien race? Were they an elite race that roamed the entire planet? Why do anthropologists rarely talk about cranial deformation and know so little about it?
250 Pages. 6x9 Paperback. Illustrated. $19.95. Code: ECD

LOST CITIES & ANCIENT MYSTERIES OF THE SOUTHWEST
By David Hatcher Childress

Join David as he starts in northern Mexico and searches for the lost mines of the Aztecs. He continues north to west Texas, delving into the mysteries of Big Bend, including mysterious Phoenician tablets discovered there and the strange lights of Marfa. Then into New Mexico where he stumbles upon a hollow mountain with a billion dollars of gold bars hidden deep inside it! In Arizona he investigates tales of Egyptian catacombs in the Grand Canyon, cruises along the Devil's Highway, and tackles the century-old mystery of the Lost Dutchman mine. In Nevada and California Childress checks out the rumors of mummified giants and weird tunnels in Death Valley, plus he searches the Mohave Desert for the mysterious remains of ancient dwellers alongside lakes that dried up tens of thousands of years ago. It's a full-tilt blast down the back roads of the Southwest in search of the weird and wondrous mysteries of the past!

486 Pages. 6x9 Paperback. Illustrated. Bibliography. $19.95. Code: LCSW

TECHNOLOGY OF THE GODS
The Incredible Sciences of the Ancients
by David Hatcher Childress

Childress looks at the technology that was allegedly used in Atlantis and the theory that the Great Pyramid of Egypt was originally a gigantic power station. He examines tales of ancient flight and the technology that it involved; how the ancients used electricity; megalithic building techniques; the use of crystal lenses and the fire from the gods; evidence of various high tech weapons in the past, including atomic weapons; ancient metallurgy and heavy machinery; the role of modern inventors such as Nikola Tesla in bringing ancient technology back into modern use; impossible artifacts; and more.

356 PAGES. 6x9 PAPERBACK. ILLUSTRATED. BIBLIOGRAPHY. $16.95. CODE: TGOD

VIMANA AIRCRAFT OF ANCIENT INDIA & ATLANTIS
by David Hatcher Childress, introduction by Ivan T. Sanderson

In this incredible volume on ancient India, authentic Indian texts such as the *Ramayana* and the *Mahabharata* are used to prove that ancient aircraft were in use more than four thousand years ago. Included in this book is the entire Fourth Century BC manuscript *Vimaanika Shastra* by the ancient author Maharishi Bharadwaaja. Also included are chapters on Atlantean technology, the incredible Rama Empire of India and the devastating wars that destroyed it.

334 PAGES. 6x9 PAPERBACK. ILLUSTRATED. $15.95. CODE: VAA

LOST CONTINENTS & THE HOLLOW EARTH
I Remember Lemuria and the Shaver Mystery
by David Hatcher Childress & Richard Shaver

Shaver's rare 1948 book *I Remember Lemuria* is reprinted in its entirety, and the book is packed with illustrations from Ray Palmer's *Amazing Stories* magazine of the 1940s. Palmer and Shaver told of tunnels running through the earth—tunnels inhabited by the Deros and Teros, humanoids from an ancient spacefaring race that had inhabited the earth, eventually going underground, hundreds of thousands of years ago. Childress discusses the famous hollow earth books and delves deep into whatever reality may be behind the stories of tunnels in the earth. Operation High Jump to Antarctica in 1947 and Admiral Byrd's bizarre statements, tunnel systems in South America and Tibet, the underground world of Agartha, the belief of UFOs coming from the South Pole, more.

344 PAGES. 6x9 PAPERBACK. ILLUSTRATED. $16.95. CODE: LCHE

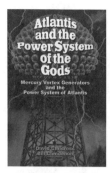

ATLANTIS & THE POWER SYSTEM OF THE GODS
by David Hatcher Childress and Bill Clendenon
Childress' fascinating analysis of Nikola Tesla's broadcast system in light of Edgar Cayce's "Terrible Crystal" and the obelisks of ancient Egypt and Ethiopia. Includes: Atlantis and its crystal power towers that broadcast energy; how these incredible power stations may still exist today; inventor Nikola Tesla's nearly identical system of power transmission; Mercury Proton Gyros and mercury vortex propulsion; more. Richly illustrated, and packed with evidence that Atlantis not only existed—it had a world-wide energy system more sophisticated than ours today.
246 PAGES. 6x9 PAPERBACK. ILLUSTRATED. $15.95. CODE: APSG

THE ANTI-GRAVITY HANDBOOK
edited by David Hatcher Childress
The new expanded compilation of material on Anti-Gravity, Free Energy, Flying Saucer Propulsion, UFOs, Suppressed Technology, NASA Cover-ups and more. Highly illustrated with patents, technical illustrations and photos. This revised and expanded edition has more material, including photos of Area 51, Nevada, the government's secret testing facility. This classic on weird science is back in a new format!
230 PAGES. 7x10 PAPERBACK. ILLUSTRATED. $16.95. CODE: AGH

ANTI–GRAVITY & THE WORLD GRID
Is the earth surrounded by an intricate electromagnetic grid network offering free energy? This compilation of material on ley lines and world power points contains chapters on the geography, mathematics, and light harmonics of the earth grid. Learn the purpose of ley lines and ancient megalithic structures located on the grid. Discover how the grid made the Philadelphia Experiment possible. Explore the Coral Castle and many other mysteries, including acoustic levitation, Tesla Shields and scalar wave weaponry. Browse through the section on anti-gravity patents, and research resources.
274 PAGES. 7x10 PAPERBACK. ILLUSTRATED. $14.95. CODE: AGW

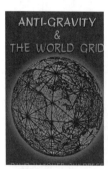

ANTI–GRAVITY & THE UNIFIED FIELD
edited by David Hatcher Childress
Is Einstein's Unified Field Theory the answer to all of our energy problems? Explored in this compilation of material is how gravity, electricity and magnetism manifest from a unified field around us. Why artificial gravity is possible; secrets of UFO propulsion; free energy; Nikola Tesla and anti-gravity airships of the 20s and 30s; flying saucers as superconducting whirls of plasma; anti-mass generators; vortex propulsion; suppressed technology; government cover-ups; gravitational pulse drive; spacecraft & more.
240 PAGES. 7x10 PAPERBACK. ILLUSTRATED. $14.95. CODE: AGU

THE TIME TRAVEL HANDBOOK
A Manual of Practical Teleportation & Time Travel
edited by David Hatcher Childress
The Time Travel Handbook takes the reader beyond the government experiments and deep into the uncharted territory of early time travellers such as Nikola Tesla and Guglielmo Marconi and their alleged time travel experiments, as well as the Wilson Brothers of EMI and their connection to the Philadelphia Experiment—the U.S. Navy's forays into invisibility, time travel, and teleportation. Childress looks into the claims of time travelling individuals, and investigates the unusual claim that the pyramids on Mars were built in the future and sent back in time. A highly visual, large format book, with patents, photos and schematics. Be the first on your block to build your own time travel device!
316 PAGES. 7x10 PAPERBACK. ILLUSTRATED. $16.95. CODE: TTH

COVERT WARS AND BREAKAWAY CIVILIZATIONS
By Joseph P. Farrell
Farrell delves into the creation of breakaway civilizations by the Nazis in South America and other parts of the world. He discusses the advanced technology that they took with them at the end of the war and the psychological war that they waged for decades on America and NATO. He investigates the secret space programs currently sponsored by the breakaway civilizations and the current militaries in control of planet Earth. Plenty of astounding accounts, documents and speculation on the incredible alternative history of hidden conflicts and secret space programs that began when World War II officially "ended."
292 Pages. 6x9 Paperback. Illustrated. $19.95. Code: BCCW

PRODIGAL GENIUS
The Life of Nikola Tesla
by John J. O'Neill
This special edition of O'Neill's book has many rare photographs of Tesla and his most advanced inventions. Tesla's eccentric personality gives his life story a strange romantic quality. He made his first million before he was forty, yet gave up his royalties in a gesture of friendship, and died almost in poverty. Tesla could see an invention in 3-D, from every angle, within his mind, before it was built; how he refused to accept the Nobel Prize; his friendships with Mark Twain, George Westinghouse and competition with Thomas Edison. Tesla is revealed as a figure of genius whose influence on the world reaches into the far future. Deluxe, illustrated edition.
408 pages. 6x9 Paperback. Illustrated. Bibliography. $18.95. Code: PRG

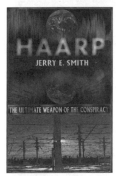

HAARP
The Ultimate Weapon of the Conspiracy
by Jerry Smith
The HAARP project in Alaska is one of the most controversial projects ever undertaken by the U.S. Government. At at worst, HAARP could be the most dangerous device ever created, a futuristic technology that is everything from super-beam weapon to world-wide mind control device. Topics include Over-the-Horizon Radar and HAARP, Mind Control, ELF and HAARP, The Telsa Connection, The Russian Woodpecker, GWEN & HAARP, Earth Penetrating Tomography, Weather Modification, Secret Science of the Conspiracy, more. Includes the complete 1987 Eastlund patent for his pulsed super-weapon that he claims was stolen by the HAARP Project.
256 pages. 6x9 Paperback. Illustrated. Bib. $14.95. Code: HARP

WEATHER WARFARE
The Military's Plan to Draft Mother Nature
by Jerry E. Smith
Weather modification in the form of cloud seeding to increase snow packs in the Sierras or suppress hail over Kansas is now an everyday affair. Underground nuclear tests in Nevada have set off earthquakes. A Russian company has been offering to sell typhoons (hurricanes) on demand since the 1990s. Scientists have been searching for ways to move hurricanes for over fifty years. In the same amount of time we went from the Wright Brothers to Neil Armstrong. Hundreds of environmental and weather modifying technologies have been patented in the United States alone – and hundreds more are being developed in civilian, academic, military and quasi-military laboratories around the world *at this moment!* Numerous ongoing military programs do inject aerosols at high altitude for communications and surveillance operations.
304 Pages. 6x9 Paperback. Illustrated. Bib. $18.95. Code: WWAR

ORDER FORM

**10% Discount
When You Order
3 or More Items!**

One Adventure Place
P.O. Box 74
Kempton, Illinois 60946
United States of America
Tel.: 815-253-6390 • Fax: 815-253-6300
Email: auphq@frontiernet.net
http://www.adventuresunlimitedpress.com

ORDERING INSTRUCTIONS

✓ Remit by USD$ Check, Money Order or Credit Card

✓ Visa, Master Card, Discover & AmEx Accepted

✓ Paypal Payments Can Be Made To:

   info@wexclub.com

✓ Prices May Change Without Notice

✓ 10% Discount for 3 or More Items

SHIPPING CHARGES

United States

✓ Postal Book Rate { $4.50 First Item
                     50¢ Each Additional Item

✓ POSTAL BOOK RATE Cannot Be Tracked!
   Not responsible for non-delivery.

✓ Priority Mail { $6.00 First Item
                  $2.00 Each Additional Item

✓ UPS { $7.00 First Item
        $1.50 Each Additional Item

   NOTE: UPS Delivery Available to Mainland USA Only

Canada

✓ Postal Air Mail { $15.00 First Item
                    $2.50 Each Additional Item

✓ Personal Checks or Bank Drafts MUST BE

   US$ and Drawn on a US Bank

✓ Canadian Postal Money Orders OK

✓ Payment MUST BE US$

All Other Countries

✓ Sorry, No Surface Delivery!

✓ Postal Air Mail { $19.00 First Item
                    $6.00 Each Additional Item

✓ Checks and Money Orders MUST BE US$
   and Drawn on a US Bank or branch.

✓ Paypal Payments Can Be Made in US$ To:
   info@wexclub.com

SPECIAL NOTES

✓ RETAILERS: Standard Discounts Available

✓ BACKORDERS: We Backorder all Out-of-
   Stock Items Unless Otherwise Requested

✓ PRO FORMA INVOICES: Available on Request

✓ DVD Return Policy: Replace defective DVDs only

ORDER ONLINE AT: www.adventuresunlimitedpress.com

**10% Discount When You Order
3 or More Items!**

Please check: ✓

☐ This is my first order	☐ I have ordered before

Name	
Address	
City	

State/Province	Postal Code

Country		
Phone: Day		Evening
Fax	Email	

Item Code	Item Description	Qty	Total

Please check: ✓

	Subtotal ▶	
	Less Discount-10% for 3 or more items ▶	
☐ Postal-Surface	Balance ▶	
☐ Postal-Air Mail	Illinois Residents 6.25% Sales Tax ▶	
(Priority in USA)	Previous Credit ▶	
☐ UPS	Shipping ▶	
(Mainland USA only)	Total (check/MO in USD$ only) ▶	

☐ Visa/MasterCard/Discover/American Express

Card Number:

Expiration Date: Security Code:

✓ SEND A CATALOG TO A FRIEND: